David Hyder

The Determinate World

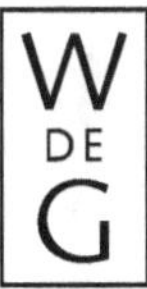

Quellen und Studien zur Philosophie

Herausgegeben von
Jens Halfwassen, Jürgen Mittelstraß,
Dominik Perler

Band 69

Walter de Gruyter · Berlin · New York

The Determinate World

Kant and Helmholtz on the Physical Meaning of Geometry

by
David Hyder

Walter de Gruyter · Berlin · New York

♾ Printed on acid-free paper which falls within
the guidelines of the ANSI to ensure permanence and durability.

ISBN 978-3-11-048157-0
ISSN 0344-8142

Library of Congress Cataloging-in-Publication Data

Hyder, David Jalal, 1964–
The determinate world : Kant and Helmholtz on the physical meaning of geometry / by David Hyder.
p. cm. – (Quellen und Studien zur Philosophie, ISSN 0344-8142 ; Bd. 69)
Includes bibliographical references and index.
ISBN 978-3-11-018391-7 (hardcover : alk. paper)
1. Geometry – Philosophy. 2. Kant, Immanuel, 1724–1894. 3. Helmholtz, Hermann von, 1821–1894. I. Title.
QA442.H93 2009
516.001–dc22

2009036201

Bibliographic information published by the Deutsche Nationalbibliothek

The Deutsche Nationalbibliothek lists this publication in the Deutsche Nationalbibliografie; detailed bibliographic data are available in the Internet at http://dnb.d-nb.de.

Printed in Germany
Cover design: Christopher Schneider, Laufen.
Printing and binding: Hubert & Co., Göttingen

Acknowledgments

I first became interested in Helmholtz while visiting the University of Göttingen in 1993 to study with Lorenz Krüger, shortly before his untimely death. I spent the years from 1997 to 2000 at the Max Planck Institute for the History and Philosophy of Science in Berlin, which he was instrumental in founding, and where I held the fellowship named after him. So this book is very much due to him. At the Institute, I profited from discussions with Claude Debru, Alexandre Métraux, Hans-Jörg Rheinberger, Jutta Schickore, Renate Wahsner and Scott Walter, as well as many colleagues in Prof. Rheinberger's departmental colloquium. In the same period, I learned much about Helmholtz and nineteenth-century German philosophy of science from Michael Heidelberger and Gregor Schiemann at the Humboldt-Universität. Discussions with Heinz Lübbig on metrology and physics gave me insight into Helmholtz's theory of measurement. During a stay at the Department of History and Philosophy of Science at the University of Indiana, Bloomington in 1998, I learned the basics of Kant's *Metaphysical Foundations of Natural Science* from Michael Friedman, whose influence on this book is evident. I also profited from the work of Daniel Sutherland on Kant's theory of magnitudes.

A great debt is owed to Konstantin Pollok, whose commentary on Kant's *Metaphysical Foundations* was invaluable, but who also helped me as a friend with countless interpretational questions. Olivier Darrigol's work on Helmhholtz provided a basis for much of my thinking on the physical ramifications of the latter's philosophy of science. Ulrich Majer's knowledge of turn of the century German philosophy of physics has always proved a valuable resource. Richard Arthur responded generously to my inquiries regarding Newton's early parallelogram proof.

During my time teaching in Konstanz 2000–2004, I profited from discussions with Bernhard Thöle and my other colleagues at the Department of Philosophy and the Centre for the Philosophy of Science. Jürgen Mittelstraß gave me the means and the impetus to write my habilitation thesis, on which this book is based, and I owe many thanks both to him and to the present editors of the series for putting it into print. I am indebted to Uta Matthies for her help and her friendship during much of my time in Germany.

Most of the research making up this book was funded by the Social Science and Humanities Research Council of Canada and the Max Planck Institute for the History of Science, to whom I express my gratitude. It was presented in many places, but I would like to thank above all the organisers and participants of the conference, "The Interaction between Mathematics, Physics and Philosophy from 1850 to 1940" at the Carlsberg Academia in Copenhagen in 2002, and participants at the Colloquium of the Department of Philosophy at the University of Marburg in 2003, above all Peter Janich. A shorter version of the arguments of Chapter 3, entitled "Kant, Helmholtz and the Determinacy of Physical Theory," was first presented in the 2007 volume resulting from the Copenhagen conference: *Interactions: Mathematics, Physics and Philosophy from 1860 to 1930*, edited by V. Hendricks, K. Jørgensen, J. Lützen and S. Pedersen. Dordrecht: Kluwer. These ideas were further developed during a stay at the Institut d'Histoire et de Philosophie des Sciences et des Techniques in Paris in 2003, and I am grateful to the institute and my colleagues there, above all Jacques Dubucs, for supporting my research.

I am indebted to Christoph Schirmer, who was a great help in the final editing of the book, and to Cynthia and Robert Swanson, for preparing the index. Special thanks are due to Gillian Grant for enduring me during the long months of editing. I am grateful for her patience and companionship.

Contents

1. Introduction

In the decades about the turn of the twentieth century, few topics were more popular among scientifically inclined philosophers and philosophically inclined scientists than the Helmholtz-Lie *Raumproblem*, or problem of space. In its philosophical guise, this problem challenged neo-Kantian philosophers and scientists to account for the advent of non-Euclidean geometries, and to explain whether and how Euclidean geometry could be said to be privileged. How could one prove that the space of intuition was Euclidean, as was generally assumed, and not pseudo-spherical? What grounds would such a proof require? One is also hard put to find a popular treatise on the subject which fails to consider the analogy between space proper and "spaces" of sensations, that is, the possibility of organising colours, tones, or tactilia in manifolds, whose spatial structure derives from the additive properties of primitive sensibilia.[1] The favoured example is inevitably that of colorimetry, and this with good reason: it was the only one of these phenomenological spaces to have been mapped out in any detail. Indeed, it is fair to say that the successes achieved in this one case by Helmholtz, Maxwell, Hering and others held out the promise of similar analyses of the other senses. Even though this promise was never truly fulfilled, this detail did little to detract from the charm of the notion. For what epistemological proposition could be more enticing to such men of science than the suggestion that the entire field of their perceptions was structured like a well-ruled graph?

To take just one example, in his *Philosophie der Mathematik und Naturwissenschaft*, Hermann Weyl connected his discussion of the various possible geometries to that of relativity theory with short excurses on Hermann Graßmann's vector calculus, and on the application of projective geometries to the continuum of colours. He concluded his discussion by remarking that, "It is important for the philosophical discussion of geometry that there exists along with space another, quite different domain of givens—the colours—which form a continuum allowing of geometri-

1 Examples of better-known authors would include (Carnap 1922; Poincaré 1909; Russell 1996; Weyl 1927).

cal treatment."[2] The modern reader may well suppose that for such authors, the notion of a colour-geometry was a mere diversion. But Weyl most certainly knew that the connection between the *Raumproblem* and colorimetry had been there from the beginning, for Helmholtz had done critical work in both areas.[3] His philosophical arguments concerning geometry rest on detailed empirical research into the nature of manifolds, where physical space realises only one among many possibilities.

In this book, I combine an analysis of Helmholtz's scientific work with an investigation of its philosophical background in order to develop a new interpretation of his work on geometry. I believe that the problems he encountered in his actual research drive much of his philosophy. But I am equally interested in Helmholtz's acknowledged Kantian heritage. In contrast to earlier authors,[4] I read Helmholtz not so much as an opponent of Kant, but rather, as Helmholtz himself often maintained, as a physicist who sought to correct internal difficulties in Kant's theory of science, without, however, abandoning its transcendental core. But this connection to Kant should not be sought on the level of general epistemology. It is true enough that Helmholtz, like many German physicists of his time, held theses of a generally Kantian stripe. Being a Kantian in this sense involves: (a) commitment to the notion of *a priori* knowledge, and perhaps to a division of the latter's sources into categories and intuitions, (b) denial of the possibility of knowledge of things in themselves. These are general Kantian epistemological theses. At the outset, Helmholtz is a Kantian in this sense, as are indeed many of his contemporaries, among them his student Hertz.[5] And although he systematically disman-

2 (Weyl 1927), p. 56.

3 The same cannot be said for Carnap. In a footnote to his discussion of the spatial arrangement of colours in his thesis, he claims that such an "application of formal space to non-spatial objects is absent from the literature" (Carnap 1922), p. 80. This remark is hard to square with the long list of Helmholtz's works appearing in his bibliography.

4 A notable exception is (Heimann 1974). While I agree with Heimann that Kant's *Metaphysical Foundations of Natural Science* exerted considerable influence on Helmholtz at this stage, I am not convinced that he succeeds in identifying the actual connection. As pointed out in (Schiemann 1997), p. 189, the analogy he proposes between the structure of the *Metaphysical Foundations* and that of Helmholtz's "Introduction" to the *Conservation of Energy* seems forced. I hope that my analysis in Chapters 2 and 3 pays stricter attention to Helmholtz's use of Kant's "constructive" theory of magnitudinal concepts and to the latter's notion of logical determination.

5 *Cf.* (Hyder 2003).

tles a large part of what Kant counted as *a priori* knowledge over the course of his career, he remains committed to (b). But this sense of being a Kantian has little immediate connection to scientific and physical problems. A stronger sense, more directly related to the philosophy of science, would involve: (c) commitment to some notion of schematisation—i. e. of the connection between pure categories (such as causality) and pure intuitions, thus also (d) the idea that the concept of matter has certain essential characteristics, and that these result from the schemata of pure concepts. Finally it involves, (e) the thesis that the schemata of "pure empirical" concepts, such as that of matter, provide what Kant calls "general laws of nature." These doctrines are specific to Kant's philosophy of *science*, and they receive their most detailed development in the *Metaphysical Foundations of Natural Science.*[6]

In the second chapter of this book, I will examine Kant's attempts to determine these general laws of nature by constructing in intuition those basic laws of representation which he calls, in the first *Critique*, the Principles of Pure Understanding. This rather technical discussion will then serve as a background for an analysis of Helmholtz's Kantian argumentation in the Introduction of his 1847 *On the Conservation of Energy.* This analysis, which forms the subject of Chapter 3, seeks the links between the theories of the philosopher and the physicist in their respective uses of regulative and constitutive principles in the sciences. But such a general connection must of course result in specific theses. I will claim that both Kant and Helmholtz postulate that there can only be central forces in nature (that is to say, in scientific *descriptions* of nature) because of their shared belief in the empirical indeterminacy of absolute space. This thesis will then provide the basis for the argument I advance in the subsequent chapters concerning the significance of Helmholz's empiricisation of geometry. This concern with the empirical status of geometry is, on my reading, not solely a reaction to the advent of non-Euclidean geometry, nor is it a product of Helmholtz's naturalism. It derives just as well from this early concern of Helmholtz's, namely that absolute space, because it is unobservable, cannot found the geometrical relations required in *physics.* This concern is essentially that expressed many years later by

6 (Kant 1911a). This and other texts of Kant will in the following be referenced by the volume and page number of the Akademie-Ausgabe, following the standard convention, i. e. 4.123 (volume 4, p. 123). References to the *Critique of Pure Reason* (Kant 1998) will use the paginations of the A and B editions, in accordance with standard practice.

Ernst Mach in the following terms: "if the inconceivable hypotheses of absolute space and absolute time cannot be accepted, the question arises: In what way can we give a comprehensible meaning to the law of inertia?"[7]

The connection between central force laws and the indeterminacy of absolute space is by no means obvious, and it will require considerable demonstration. My strategy involves showing how Kant's appeal to central forces forms an essential part of his *a priori* derivation of the various laws of matter and motion that are presented in the *Metaphysical Foundations.* For instance, Kant attempts to give an *a priori* proof of Newton's parallelogram law of force composition by appealing to the relativity of spatial relations. According to Kant, in order for us to imagine that a body is undergoing two motions simultaneously, we must be able to refer each of these motions to at least one other point. Thus the construction of two instantaneous velocity-changes which sum to form a third requires that we imagine three points: two of them are the causes of changes in the motion of the third, such that each change is relative to one of the two points; and the total motion (or total acceleration, in the instantaneous case) is then the motion of the point relative to the centre of mass of the system. It then follows that the actions of the pairs of points on each other sum geometrically to yield the action of each individual point with respect to the centre of mass. Thus forces *must* sum geometrically, if the motions they cause can be determinately constructed in intuition. This (fallacious) proof is supposedly superior to Newton's demonstration of the parallelogram law not only because it is apodictic, but also because it involves no appeals to absolute space. Indeed, it appeals to the *empirical indeterminacy* of space (its emptiness) to justify the claim that only centrally directed changes in motion are conceivable. This sort of transcendental deduction of physical principles typifies Kant's method in the *Metaphysical Foundations*—a method which Helmholtz follows quite faithfully in the *Conservation of Energy,* and in an early manuscript from this period on *a priori* principles in the sciences.

In Chapter 3, I will argue that Helmholtz borrows this argument of Kant not least because he views the notion of absolute space with equal, if not greater suspicion. By Helmholtz's time, however, new work in electrodynamics had rendered the status of absolute space genuinely problematic. Helmholtz opposes electrodynamic theories which involve forces that depend on the motions and accelerations of electric

7 (Mach 1960), p. 293

quanta, because such dependencies threaten to introduce empirically indeterminate quantities into physical theory. By this he means, as he still insisted almost fifty years later, that an epistemologically legitimate theory should involve only forces that "must necessarily be determined once the positions of the masses are known."[8] Helmholtz consistently maintained, in agreement with Kant, that the fundamental magnitudes that determine the evolution of a system must be restricted to the masses and positions of the particles in the system (forces are here viewed as essential properties of kinds of matter, whose intensity varies with relative position). And the positions of the masses in question can only be relative positions, since absolute positions are not possible experiences. Like Kant, Helmholtz takes these epistemological theses to imply the necessity of central forces, and thereby to disqualify theories such as those of Weber and Neumann, as well as the later theory of Maxwell, in which forces that do not depend on relative position alone are admitted.[9] He reserves special opprobrium for the last sort of theory, in which "a force is made dependent on absolute motion, that is to say on a changing relation between a mass and something that can never be an object of a possible experience."[10]

In the concluding section of Chapter 3, I will suggest that criticisms directed against Helmholtz by the physicist Rudolph Clausius[11] led Helmholtz to address explicitly the distinction between what he called purely mathematical coordinate systems, and ones which were physically significant. Although he intended here to distinguish only between relative and absolute directions, I shall show that he was in fact appealing to preliminary investigations into the nature of geometrical magnitudes that he had undertaken before the publication of the *Conservation.* These early manuscripts, which shall be discussed in Chapter 5, provide crucial evidence for my claim that Helmholtz was already confronting the central flaw in the Kantian system that he had borrowed. The problem was that while Kant had appealed repeatedly to the indeterminate character of space in order to transcendentally deduce the necessary form of certain

8 This quotation and the following one are taken from Helmholtz's 1882 comments on the original edition of the *Conservation.* They therefore provide valuable insight into the evolution of his thought over the intervening forty-five years. (Helmholtz 1996), p. 54.

9 *Cf.* (Darrigol 1994), pp. 225–227 for a discussion of Helmholtz's objections to Weber's and Neumann's theories.

10 (Helmholtz 1996), p. 55.

11 (Clausius 1853, 1854)

physical laws, he had never problematised the sense in which *geometrical* propositions count as "determinate." Recalling Mach's remark from above, we might say that Kant requires geometry to be determinate in order that kinematics may be so. But the determinacy of geometry is not compatible with the claim that absolute space is not a possible object of experience. It is, I shall be suggesting, this problem that provided an initial trigger to Helmholtz's geometrical investigations of the 1860's and 1870's.

Before examining that development in Chapters 5 and 6, I will consider in Chapter 4 the relation I mentioned, by way of introduction, between Helmholtz's work on colour-perception and the problem of space. Here again, my analysis differs from the conventional reading,[12] according to which Helmholtz was inspired by his physiological work because it provided him with evidence for the *inductive* origins of metrical relations. On my view, it is not the naturalist's concern with subliminal induction that was decisive here. Rather, I suggest that Helmholtz learned specific lessons concerning the role of operationally realisable measurement axioms in the determination of a metric space. This insistence on operational realisation is what distinguishes his work on geometry from the mathematically more sophisticated work of Riemann.[13] Helmholtz himself emphasised the connections between colour-theory and geometry, pointing out that when he first began his geometrical investigations "two examples were available to me in physiological optics itself of other severally variable [*mehrfach veränderliche*] manifolds which can be portrayed spatially; namely, the system of colours, which Riemann cites, and the measuring out of the visual field by the eye."[14] It is the connection between measurement operations and spatial metrics which was the crux of his later works on physical geometry.

The origins of this colour-research have no obvious connections to problems in the theory of geometry, and there is no suggestion that Helmholtz was aware of such a possible connection at the outset. In 1852, he began a research programme aimed at defending Newton's theory of colour-perception against those advanced in the intervening period by Mayer, Brewster and others. The Newtonian theory argued that colours were of two sorts: pure spectral colours, whose causes were the "un-

12 See for instance (Lenoir 1993; DiSalle 1993), who do follow up on Helmholtz's reference in the geometry papers to his sense-physiological research.

13 (Riemann 1854)

14 (Helmholtz 1868b), p. 619. Helmholtz's reference is to (Riemann 1854).

mixed" spectral lights that resulted from decomposing white light by means of a prism; and "mixed" colours, which resulted from the combined action of spectral lights. Newton presented a device for calculating the shades of mixed colours that would result when different quantities of spectral lights acted simultaneously on the eye. This device—Newton's colour-wheel—was a circular diagram with the spectral colours arrayed on the circumference. The colour resulting from a mixture could be determined by taking the "weighted sum" of the colours mixed: one drew a line between the two colours on the circumference, selecting that point which divided the line in proportion to the quantities of the two mixed colours. The colour of that point was given by extending a line from the centre of the circle through the point to a spectral colour on the circumference. Its intensity was proportional to the distance of the point from the centre. Thus the colours were arranged on a wheel, with white at the centre, the pure spectral colours at the circumference, and the intermediate colours arranged in between in concentric shells of equal intensity. In a weak sense, colours formed a space, which involved rudimentary notions of *line* and *distance.*

In the next five years, Helmholtz's research was criticised by the mathematician Hermann Graßmann, whose work was in turn adapted and extended by the young James Clerk Maxwell. Graßmann, the inventor of vector algebra, reinterpreted Newton's embryonic model as a vector-space, in which each spectral colour was treated as a vector, and the mixed colours were then given by vector additions. More importantly, he cast the relations among the various colours in axiomatic form, emphasising that each of these axioms had to be correlated with a corresponding definition or operation: for instance, the axiom of vector addition had to be correlated with a specific mixture procedure. In the course of analysing Graßmann's formal extension of his research, Helmholtz was first confronted with the possibility that what counts as a "distance" within the continuous manifold of colours was partly a result of the axioms used to characterise the space, partly the result of the measurement operation chosen, and, of course, partly the result of performing it. Distance was in other words not an inherent property of the manifold, but one that was induced by operational factors.

There are further connections between Helmholtz's sense-physiological research and his later papers of geometry. For instance, in the last of his papers on the subject, "Über den Ursprung und Sinn der geometrischen Sätze; Antwort gegen Herrn Professor Land," Helmholtz argues against a neo-Kantian opponent in a purely "idealistic" voice, asking the latter to concede that appearances in space must have two sort of causes, "topogeneous" and "hylogeneous moments." When we talk about bodies moving about in space, we are always talking about an interplay between both sorts of moments or causes: that is to say, any description of the behaviour of material bodies in space involves reference to connections between material properties, such as colours and tactilia, and spatial locations. An image on the retina is composed of stimulations of colour receptors at particular points (Lotzian "*Lokalzeichen*"), and a moving image may, or may not, preserve its internal relations as the eye sweeps its field. Similarly the combinations of material and spatial properties making up a *Raumgebilde*, or spatial form, may or may not preserve their structure as we move them (they might, after all, simply disappear, the object simply evaporate the moment we touched it). But while it is true that this sense-physiological background gives us a partial insight into the sorts of models Helmholtz had in mind when proposing methods for "visualising" non-Euclidean geometry, it does not get to the origin of Helmholtz's most important insight, which concerned the interrelation of intuitions, measurement and axiomatic systems. It was in this regard that "colour-geometry," to use Weyl's phrase, played a critical role. To make this point clear, we can begin by considering the various possible interpretations one can offer of Helmholtz's four seminal papers in geometry. For even a cursory consideration of the arguments in his geometrical papers leaves the reader wondering *in what sense* Helmholtz thinks geometry is an empirical science. In Chapters 5 and 6, I will attempt to give a detailed answer to this question.

There are several possible alternatives. Most simply, his position might be that space can be "curved" in many different ways. We cannot be sure what this intrinsic curvature might be without carrying out empirical measurements. A more sophisticated argument would be that, although a number of geometries are consistent with the space of our experience, only one can be accommodated to physical theory. A further possibility is that geometry is itself a kind of physical theory, in that it describes not the properties of space itself, but only those of bodies in space. Finally, Helmholtz's point may be that geometry is neither a physical science, nor a science concerned with space itself, but rather a science of measurement, which is empirical because it is involved with the material systems used in measurement. I will be urging the last interpretation in what follows. However it must be conceded that Helmholtz does not systematically distinguish between the above alternatives, and that he makes numerous statements which, on their own, support each of them. So we must begin by looking at the main lines of the four papers in question.

Here we find two quite distinct arguments: a positive "transcendental"[15] one, which deals with the physical bases of measurement systems; and a negative one concerning the intuitive imaginability of alternative geometries. The first of these is advanced in his first two papers on geometry from 1868.[16] The second argument is the main concern of his third, 1870 paper, and the last one, from 1878, combines the two.[17] This development is largely the result of Helmholtz's overlooking, in 1868, the possibility of a pseudo-spherical geometry. In the two papers of that year, entitled "Über die tatsächlichen Grundlagen der Geometrie" and "Über die Tatsachen, die der Geometrie zum Grunde Liegen,"[18] Helmholtz had argued that in applying geometry to experience, we assume that it *makes sense* to speak of relations of congruence between arbitrarily distant, and arbitrarily situated forms in space. Axioms and theorems speak of line segments of equal, greater and lesser length, and postulates enjoin us to construct the latter. They thereby presuppose that we can ascertain whether two line segments stand in these relations. This supposition entails that it be possible to compare such forms with one another, and therefore that one can make such comparisons by means of translating and rotating rigid bodies. In consequence, geometry presupposes a factual basis, namely that bodies and motions suited to the measurement-operations presupposed by the axioms are available.

The arguments of these two papers differed only slightly, since the one was the text of a public lecture in which Helmholtz announced and summarised the results of the longer, technical publication. In the latter,[19] Helmholtz derived a unique expression defining distance in those metrics compatible with these conditions, and showed that this expression agreed with that of Riemann for the Euclidean metric. Shortly after announcing this result, however, he realised that his analysis did not disallow pseudo-spherical metrics. He concluded that geometry was empirical in a second, stronger sense: Euclidean and pseudo-spherical ge-

---

15 (Torretti 1978), p. 168.
16 (Helmholtz 1868a, 1868b)
17 (Helmholtz 1870, 1878c)
18 The first paper (Helmholtz 1868a) is falsely dated 1866 in Helmholtz's *Wissenschaftliche Abhandlungen.* This error is generally reproduced in the literature, and sometimes leads commentators to overlook the fact that the argumentation of these two papers is quite distinct from that of the subsequent two.
19 (Helmholtz 1868b)

ometries disagree with regard to the behaviours of rigid bodies; alternatively, for the same phenomena, they disagree on what counts as a rigid body. Thus there is an empirical question as to which of the two geometries is valid. Helmholtz's position thereby underwent a radical shift, which he only partially acknowledged. In the papers of 1870 and 1878, "Über den Ursprung und Bedeutung der geometrischen Axiome" and "Über den Ursprung und Sinn der geometrischen Sätze,"[20] he presented a number of by now well-known thought experiments (two-dimensional beings living on the surface of a sphere, the spatial relations in the world of a convex mirror) which were intended to demonstrate the compossibility of Euclidean and non-Euclidean spatial metrics. This shows that there is no logical or intuitive impediment to imagining movements of rigid bodies, or, in the 1878 paper, "physically equivalent magnitudes," which agree to a pseudo-spherical metric. Since either sort of metric is possible, there is an empirical question as to which applies.

This later argument obviously builds on the earlier one from 1868. In the 1868 papers, Helmholtz maintains that geometrical propositions *have meaning* if and only if certain contingent facts obtain. In the second, he argues that when two competing geometrical systems depend on distinct sets of such facts, asking which geometry is valid will be equivalent to asking which set of facts is the case. The first step has been called a transcendental one, because it argues that if (mathematical) language concerning distances in space is to make sense, then certain possibilities must be realised. The second seems more straightforward: if neither logic nor intuition can decide between two exclusive possibilities, then it is a contingent and empirical question as to which obtains.

There are two fundamental objections to Helmholtz's claims. First, as Helmholtz himself acknowledged, it is hard to see how we could know whether the facts at the basis of geometry obtained without assuming it in the first place. If we had to determine whether or not a body that had been rotated was rigid, we could not do so without assuming that the question made sense in the first place. Practically, this means that the mere coincidence of two bodies at various points in space does not entitle us to call them rigid; furthermore, if we *do* choose to regard them as rigid, we are tacitly appealing to the idea that they have not suffered change due to forces, and thus we must have some notion of lengths underlying, and independent of, the bodies in question. So "rigidity" depends on "equal length" conceptually, not the other way round. This calls

20 (Helmholtz 1870, 1878c)

the first of the arguments in to question.[21] Second, any set of bodies whose behaviours we take as supporting one or the other geometrical candidate can, under suitable transformations of mechanical laws, be redescribed as non-rigid, or as subject to the actions of forces. This criticism, which Helmholtz also anticipated, is targeted at the second argument. Taken together, these two objections undergird Poincaré's conventionalist critique. Given that the objections are good ones, and that Helmholtz himself had raised them both, why did he not see them as calling his project into question?

A common answer to this question is that Helmholtz was hindered from seeing their pertinence by his commitment to a physiological conception of space.[22] Because he failed to grasp the constitutive role of spatial intuition in Kant's epistemology, he assimilated the latter's remarks concerning the necessity of geometrical knowledge to claims about the character of subjective experience.[23] On this reading, it is the alternative visualisations which do the work for Helmholtz, for they were to demonstrate that Kant was wrong in thinking that empirical intuitions had to conform to Euclidean geometry. This deficiency in Helmholtz's reasoning is thus reflected in both of the arguments described above. For in both cases, Helmholtz mistakenly thinks that he can establish concepts like "rigid body" or "physically equivalent magnitudes" without appealing to an *a priori* framework. A different line is taken by Gregor Schiemann,[24] who has argued recently that Helmholtz's investigations into geometry are fundamentally determined by a "mechanistic world-view." On Schiemann's account, Helmholtz is blind to the second of the above-mentioned objections because he believes that geometrical properties can be derived from mechanical ones.[25] Since, however, one cannot appeal to prior knowledge of mechanical laws in order to select an appropriate geometry, the claim that one or the other geometry is empirically valid cannot be maintained.

As there are numerous passages to support both critiques, it would be senseless to argue that Helmholtz is innocent of these charges. Nevertheless, both the psychologistic and mechanistic readings pay too little atten-

21 *Cf.* (Wahsner 1994), pp. 250–251.
22 (Hatfield 1990), p. 220 ff., (DiSalle 1993), p. 514.
23 (Hatfield 1990), pp. 222–224, (DiSalle 1993), pp. 505–506.
24 (Schiemann 1997), pp. 219 ff.
25 "Instead of recognising that one cannot derive the Euclidean character of real space from the validity of the mechanical law of inertia, [Helmholtz] commits just this fallacy." (Schiemann 1997), p. 233.

tion to the centrality of measurement in Helmholtz's papers.[26] For Helmholtz, his professional interest in the physiology of the senses notwithstanding, characterises geometry repeatedly as a science of spatial measurement. And his discussions of the relations between alternative geometries and mechanics stress geometry's role in making the quantitative description of spatial objects and events possible. How is this explicit emphasis borne out in the structure of Helmholtz's arguments?

Let us begin with the so-called transcendental argument of the first two papers. Geometry is defined there as a science of spatial measurement. But Helmholtz does not mean by this that geometrical propositions are statements about inherent properties of space. He seeks instead to complicate the relations between (1) axiomatic systems and (2) the spatial continuum, by introducing a third element (3) the measurement operation. The possibility of carrying out (3) in practice is a condition for the significant application of the propositions of (1) to the elements of (2). More precisely, the availability of (3) is the empirical condition under which measurements which satisfy the axioms of (1) can be applied to objects situated in (2). For if there were no such operations, distinct spatial objects might intuitively appear to satisfy relations such as "longer" or "shorter," or "congruent," but there would be no possibility of *verifying* such relational statements.

In "Über den Ursprung und Bedeutung der geometrischen Axiome," his popular lecture of 1870, Helmholtz emphasises the possibility of imagining both pseudo-spherical and Euclidean universes, arguing that the behaviours of the objects used for carrying out measurements (our most rigid bodies) can be consistently imagined to be either Euclidean or pseudo-spherical. Nevertheless, he admits in conclusion that a "stringent Kantian" might choose to hold to Euclidean geometry in the face of such evidence, and would thereby demand that the objects involved in measurement operations conform to requirements that derive "analytically" (purely logically) from the axioms he held to. This possibility exists because it is only once we have supplemented geometrical propositions with physical ones, such as the principle of inertia, that they gain "a real content, that can be confirmed or rejected by experience."[27] Finally, in his 1878 paper, "Über den Ursprung und Sinn der geometrischen Sätze," Helmholtz mobilises this latter insight by arguing that the sufficiency of a method of measuring must in the last instance be evaluated

26 Carrier (1994) and Wahsner (1994) address this topic in greater detail.
27 (Helmholtz 1870), p. 30.

relative to a concept of "physically equivalent magnitudes" appealed to in physics. Physically equivalent magnitudes are ones in which the same processes can exist or take place in equal periods of time. In contrast to the argument of the first two papers, which claimed that Euclidean geometry was the unique logical consequence of requirements placed on measurement,[28] the later argument departs from the assumption that measurement must define and quantify those spatial magnitudes that are presupposed by physical laws. In both cases, however, Helmholtz's claim is that in order for certain metrological concepts, such as "equal length" or "physically equivalent magnitude," to have a determinate meaning, a measurement procedure must be at hand.

In sum, Helmholtz's papers on geometry go through two phases. At first convinced that the definitions of rigidity, free mobility, and rotation were sufficient to determine a unique metric, Helmholtz argued that (1) the truth of Euclidean geometry was equivalent to (2) the meaningfulness of relational statements concerning spatial measurements. And since (2) was equivalent to (3) the physical realisation of the definition of a freely movable rigid body, it followed that (1) and (3) implied one another. Euclidean geometry is true because a particular kind of measurement *must be* carried out in the spatial continuum. The "must" that is in question here derives from our needs in doing science. It is a regulative demand. In the second phase, Helmholtz admitted that his analysis did not uniquely determine which of a family of geometries was valid. In the papers of 1870 and 1878, he emphasised the compossibility of Euclidean and pseudo-spherical measurements systems, while admitting that an absolute decision between the two could only be made in the light of mechanical principles and an (operational) definition of length with respect to time. No longer content with the claim that the application of geometry to experience presupposed certain behaviours among bodies, he now argued that the results of spatial measurements were meaningful only in so far as they could be related to propositions of physics. Since general physical laws require spatial units relating physical processes to time in order to be stated at all, it is the responsibility of geometry to codify the relations holding between physically equivalent spatial magnitudes. He termed such a geometry a physical geometry.

28 Thus that it followed, in the language of the third paper, "analytically" from these requirements.

Helmholtz gave a detailed development of this argument in his last paper on the subject.[29] The paper presents the same argument in two modes: a realistic and an idealistic one. Realistically, the question of what geometry should count as objectively valid is characterised as being equivalent to determining "the objective equivalence of the real substrates of spatial magnitudes, which [equivalence] shows itself in the unfolding of physical relations and processes [*welche sich im Ablauf physischer Verhältnisse und Vorgänge bewährt*]."[30] From the idealistic point of view, since this objective equivalence cannot be checked against underlying relations among these substrates, it can only be arrived at by means of measurement operations. Helmholtz calls those elements of (noumenal) reality that are responsible for our observing an object at a given location at a given time "topogeneous factors." These factors interact with "hylogeneous factors" (the causes of material properties) to produce appearances of movable bodies in space. Physical geometry presupposes the existence of law-like behaviours of such bodies, thus in turn regular connections between the topogeneous and hylogeneous factors.[31] From the realistic point of view, we want a physical geometry to tell us which groups of topogeneous factors give rise to physically equivalent spatial magnitudes, in that they interact with the material ("hylogeneous") factors in a law-like fashion. Such a geometry would be objectively valid in the case that the relations of equivalence it ascribed to the spatial magnitudes was preserved when mapped onto the topogeneous factors.[32] The problem is that we have no access to the topogeneous factors. If space is transcendental in the sense that we could never know its noumenal ground, then its "real" metrical structure is also unknowable. We then have no choice but to give it one by settling on a physical means of determining these equivalence relations.

How then, should we understand this terminal position? In what sense did Helmholtz take geometry to be empirical in this last paper? In the preceding discussion, I outlined two positions he might have adopted: (1) geometry is empirical in that it is a species of mechanics (Helmholtz as mechanist), or (2) geometry is empirical in that it is

29 (Helmholtz 1878c)

30 (Helmholtz 1878c), p. 641.

31 See (Hyder 1999) for an account of the structure of this argument.

32 *Cf.* (DiSalle 1993), p. 517, "So the laws which our mind represents as the structure of space are just the natural laws governing the topogeneous factors." Strictly speaking, however, the laws in question are those governing the interactions of the topogeneous and hylogeneous factors.

learned by observing the behaviours of physical objects (Helmholtz as naturalist, or physiologist). I will argue instead that Helmholtz came to the position that geometry is empirical in that (1) it does not describe inherent properties of the spatial continuum, and (2) it is objectively indeterminate until supplemented with regulative constraints. From this it followed for him that (3) an operational description of spatial measurement procedures had to specified, which would make use of rigid bodies (and clocks, although Helmholtz did not go into this), where both the concepts of rigidity and equal length of time are defined operationally, as opposed to theoretically. If the formulation of physical principles demands that concepts of length be determinate, and if it is impossible to determine length relationships *a priori*, it follows that some sort of operational basis must be *imposed* on the continuum. Thus the decision between various competing geometries will depend on the adequacy of the group in question to the demands of physics. I will argue that Helmholtz's criterion here as well was derived from a regulative demand, namely the principle that all physical laws be positionally neutral or indeterminate. The preferred geometry will be that which does not lead to laws of motion that vary with position.

In Chapter 6, as well as in the Conclusion, I will connect this theory of geometry to the project of the *Conservation* by considering a problem implicit in Kant's arguments in the *Metaphysical Foundations*—a problem which, as I suggested earlier, Helmholtz inherited. In the Introduction to his *Conservation of Energy*, Helmholtz had endorsed Kant's view that the task of natural science was to tie back the changeable in experience to changes in the position of matter in space and time. In other words, he endorsed wholeheartedly the Kantian project of codifying experience by connecting (or reducing) variations of *intensive* magnitudes to variations between *extensive* ones. According to the first *Critique*, an extensive magnitude is one in which the representation of the parts is necessary for the representations of the whole. Space and time are extensive magnitudes, and thus every empirical intuition has an extensive *form*. By virtue of their part-whole containment structure, extensive magnitudes are inherently additive, and thus inherently mathematisable.[33] Intensive magnitudes, by contrast, are magnitudes that do not include other ones as their proper parts. Such intensive magnitudes are identified with the *content* of empirical intuitions, by which Kant means sensibilia such as colours and

33 I am grateful to Daniel Sutherland for first bringing these aspects of Kant's theory of magnitudes to my attention. See for instance (Sutherland 2004, 2005).

sounds; however, in the *Metaphysical Foundations*, both velocity and force are called intensive magnitudes as well. Thus a major aim of the *Metaphysical Foundations* is to provide definitions of these intensive magnitudes that found their additive properties in extensive ones. But there are only two sets of true extensive magnitudes, namely the relations among spatial and temporal forms, i.e. the pure intuitions of space and time. Since the latter form the ground of geometry and mathematics, the extensive definitions of kinematic and dynamic concepts are essential to the project of mathematising experience, that is to say to the very possibility of theoretical physics.

Thus the Kantian theory of natural science requires that intensive magnitudes of all sorts be given a mechanical reduction to extensive relations. Typically enough for Kant, a methodological directive that previously rested on a metaphysical thesis is reformulated as a regulative demand resting on epistemology. We seek to reduce phenomena to mechanical actions not because we have reason to believe that nature consists of extended matter. Rather such a reduction is required in order that physics may become mathematical. Helmholtz's work on colour-research was part and parcel of this research programme: tying colours and their relations back to objective statements about spatio-temporal properties of light was a typical piece of research in the neo-Kantian tradition of sense-physiology as carried out by Müller, Herbart and Lotze, to name a few of Helmholtz's predecessors. In the course of this research, Helmholtz, as well as Graßmann and Maxwell, was forced to reflect on the concept of a phenomenal continuum and the conditions under which "distances" in such a continuum could be defined. This research boomeranged, for Helmholtz then used it to undermine a key assumption of the Kantian project.

The assumption in question was that whereas intensive magnitudes were not fully mathematisable, the extensive magnitudes of time and space *were.* Thus Kant had seen it as the task of what he called phoronomy to provide a "construction" of the intensive magnitude of speed in terms of the intuitions of space and time. The Phoronomy of his *Metaphysical Foundations* gives a recipe for adding and subtracting speeds by treating them as line segments;[34] in other words, it assumes that the sciences of the pure extensive magnitudes are certain and *a priori.* On my

34 In the Phoronomy of the *Metaphysical Foundations of Natural Science*, Kant represents speeds as directed line segments, and shows how this presentation presupposes a (relatively) unmoving empirical space in which a body is in motion.

reading, we should interpret Helmholtz as arguing that our pure intuitions of extensive magnitudes cannot possibly do the reductive job assigned to them. For just as the system of differences that comprises the colour-space is bereft of metrical relations until measurement conventions are imposed on it, the same holds for space itself, according to Helmholtz. These conventions can only be implemented by using material objects in space, and the relations thus singled out are in turn not objectively significant until they are connected to laws describing the motions of bodies. For Helmholtz, space has no metrical characteristics, thus no geometry at all, until such operational and theoretical demands are met.

Now, the project of the *Conservation* assumed that whatever concepts were introduced in physics, they should be physically determinate. Indeed, Helmholtz had argued against Rudolph Clausius's attempts to define force functions in terms of arbitrary directions in space by accusing Clausius of employing "purely mathematical" relations in these definitions. Such relations were, on his view, the result of employing coordinate systems drawn "on paper."[35] Whereas, Helmholtz claims, we can admit in physical theory only those relations which are determined by "real things." It is striking enough that Helmholtz employs almost exactly the same language twenty-five years later to characterise the difference between what he called "pure intuitive" geometry and his "physical geometry." In both cases, I would contend, Helmholtz is appealing to the necessity of empirically grounding spatial relations because he suspects that any *a priori* assumptions concerning the geometrical structure of space runs the risk of absolutising it. And absolute space, since it is not an object of possible experience, has no role to play in physics. Of course this criticism is not yet, in the reply to Clausius, a demand for an empirical account of geometry. I would rather suggest that Helmholtz's adherence to Kant's requirement of empirical determination, which came to a head in the debate with Clausius, forced him to recognise that geometrical propositions were *inherently* connected to the basic propositions of kinematics. Indeed, he had already taken this line in an earlier manuscript. The colour-research he was conducting at the same time offered him the means to eliminate such non-empirical determinants of space by operationalising the notion of length. Thus, when he returned to physics in the late 1860's, and thereby also confronted the problem raised by Maxwell's ether theory of electromagnetism, he attempted to develop a

35 (Helmholtz 1854), p. 84.

theory of geometry that was consistent with the transcendental argumentation of the *Conservation.* This interpretation has the virtue of explaining why Helmholtz, the ardent empiricist, presents transcendental arguments in favour of a Euclidean rigid-body geometry in the 1868 papers.

Seen from this point of view, Helmholtz's work on geometry is not accidentally related to the subsequent applications that Poincaré and Einstein made of it. If I am correct in the analysis I offer, it was motivated from the start by a similar concern: Newtonian physics presupposes the validity of geometry in the statement of the laws of motion, above all the law of inertia. But if absolute space is truly an inadmissible concept, how can we give geometrical statements a determinate meaning? How are they to be related to the basic magnitudes, for instance inertial paths, of physical theory? Helmholtz, although he was working well before the critical work of Lorentz and Poincaré, had already perceived the tension between Newton's theory of space and the onset of modern electrodynamics. His provisional solution was not suited to Einstein's borrowing by chance—it was in fact an integral part of a single research programme.

2. The Empirical Determination of Physical Concepts in Kant's *Metaphysical Foundations of Natural Science*

The main arguments of this study concern the status of geometry in Helmholtz's philosophy of science. However, as I outlined in the Introduction, Helmholtz begins to question the epistemological basis of geometric science at a point in his philosophical development when he is still a quite orthodox Kantian. This adherence to Kant's philosophy is evident enough to a careful reader of the introductory sections of his 1847 monograph, the *Conservation of Energy*. But, should there be any doubt on the matter, Helmholtz himself confirms this early commitment in his 1882 comments on the monograph, where he regrets his excessive reliance on Kant, observing that "[t]he philosophical arguments of the Introduction are influenced by epistemological views of Kant that are stronger than what I would be prepared to accept today."[1] I shall be arguing throughout that even when Helmholtz undertakes a radical critique of Kant's views of geometry, he continues to do so within a theory with a strong transcendental element. For instance, even the last paper on geometry appeals to regulative constraints on the concepts that we can admit in natural science, for the latter must always be chosen so as to permit maximally general statements of physical laws. Of course, in making this claim, I do not discharge myself of the obligation to prove it, and that demonstration will in fact be given in the succeeding chapters. But whether or not the reader accepts my conclusions, I cannot argue for them until a general characterisation of Kant's philosophy of science is in place. The purpose of this chapter is therefore to outline that philosophy in a manner that is both true to Kant and serviceable to our analysis of Helmholtz.

To some extent, these aims must conflict: I cannot hope in the following to give an analysis of Kant's work that takes full account of the critical philosophy. For the text I shall be analysing, the *Metaphysical Foundations of Natural Science*, is not concerned with the philosophy of

1 (Helmholtz 1996), p. 53.

science alone. It has a specific role to play within Kant's epistemology, and this fact plays an important role in the structure of Kant's arguments. Thus before we turn to a detailed consideration of the natural scientific content of this work, we would do well to consider its context within the argument of the first *Critique*, at least to the extent that this context impinges on the logic of the *Metaphysical Foundations*. In the following few introductory remarks, I will explain how the *Metaphysical Foundations* supplements the arguments of the Transcendental Deduction and the Schematism of the categories by extending the Principles of Pure Understanding. Kant aims to show that the resulting, extended principles form the core of physical science. After explaining this relation, I will briefly summarise the breakdown of the chapter, which summary will conclude by indicating the connection to Helmholtz's work that I draw in the concluding section.

When I say that the *Metaphysical Foundations* supplements the Transcendental Deduction, I do not mean that it fills a gap in either the A- or the B-versions of the deduction given in the *Critique*. Whether or not one accepts these proofs, their tactical purpose is achieved at the point where Kant has demonstrated that the pure operations of the understanding are necessary for the unification of the manifold of experience. The deduction appeals only to the role of these operations in synthesising sensual data, and it is not concerned with the specific categories involved in the formation of empirical cognitions. That part of the demonstration is first undertaken in the Schematism, where Kant goes on to explain how each category, whose necessity has now been generally demonstrated, can in fact be related to possible cognitions by means of its schema in the pure intuition of time. From hereon in, Kant turns to what I might call a positive demonstration. The concern is no longer to explain *why* it must be the case that we categorically determine the manifold of our experience. Kant now wishes to explain *how* exactly we go about this.

So long as one stays within the purely transcendental part of philosophy, this task is completed in the section on the Principles of Pure Understanding—the Axioms of Intuition, the Anticipations of Perception, the Analogies of Experience, and the Postulates of Empirical Thought. The *Metaphysical Foundations* represents the next step in this development, but because it is concerned with matters empirical, it no longer belongs to the transcendental philosophy *per se*. Nevertheless, as I shall now explain in more detail, its method is entirely continuous with the latter, and it makes good on a number of promises made in the *Critique* and the *Prolegomena*, but which are deferred by Kant to a later discussion. One might say that the Transcendental Deduction and the Schematism demonstrate that the categories are conditions on the possibility of experience by means of a negative, and a positive demonstration. The negative demonstration involves arguing that if there were no pure operations of the understanding, then there would be no unification of experience, and thus there would be no transcendental unity of apperception, so that we would have no conscious experience at all. Thus whatever is not unified by means of such operations is simply not a possible datum of experience. The positive demonstration involves explaining how those operations that characterise our human consciousness actually take hold of the manifold of sensory data. Taken together, these demonstrations are supposed to show that these operations are necessary (the negative part), and sufficient (the positive part) to the unification of experience.

A detailed examination of these portions of the first *Critique* would obviously lie well outside the concerns of this study. What concerns us is the next phase of Kant's argumentation, namely the notion of logical determination by means of the categories. Even if one concedes to Kant that his arguments in the Deductions and the Schematism are correct, one could rightly object that in showing that the categories are essentially involved in the unity of conscious experience, Kant has yet to show that they play the logical and metaphysical roles ascribed to them in the scholastic tradition. Aristotle's position was that the categories, because they describe the fundamental structure of being, are also the fundamental structures of propositions that assert things about beings. And thus they are also the fundamental structures involved in the formulation of logical and natural laws. Since Kant is fundamentally committed to denying that the categories describe the essential properties of things in themselves, he cannot accept a correspondence of this sort. But he does want to maintain the logical doctrine: the validity of logic still derives from the role of the categories in constituting experience. And thus the transcendental argument does not stop with the claim that the categories *unify* experience. They are also essentially involved in what Kant calls the *determination* of experience.

Kant's working hypothesis is borrowed from the Leibniz-Wolffian tradition. For Leibniz, we may recall, a complete description of the concept of a monad would involve a list of all its properties. To say that a property holds of a thing or that it does not, is to *determine* the thing with respect to that property. An infinite intelligence would know the concept of every monad, and thus it would be able to say of each possible property *whether or not* the monad had it. Finite intellects, such as those of human beings, can only hope to achieve a part of this knowledge: their determinations of objects and concepts are incomplete; however, when they engage in scientific investigations, their aim is to achieve as complete a determination of objects and concepts as is possible. Once Kant has taken his Copernican turn, the entire metaphysical side of this theory is bracketed. There is no metaphysical guarantee that a successive determination of empirical concepts can or will converge on the divine state of knowledge. Nevertheless, the notion of a complete determination of empirical concepts is preserved in what Kant describes in the Architectonic of Pure Reason.[2] Reason sets itself the goal of establishing a complete system of nature, in which the manifold of concepts is unified under an idea. But such a completed system of natural laws and concepts remains an ideal towards which we strive. There can be no metaphysical proof, for Kant, that reality has an underlying structure which could, even in principle, ensure that such a system is possible.

If one fuses the Leibniz-Wolffian model with the Aristotelian account of the categories *without* taking the transcendental turn, the role of the categories is the following. The categories, since they are the fundamental metaphysical modes (the fundamental sets of *kinds* of properties), are also the fundamental logical modes. Whatever sort of determination we may undertake in our thinking, we know that it is going to play out within this categorical structure. To take a simple example: We may not know what colour to predicate of any given physical body. But, if we know that it is a physical body, we can say with absolute certainty that it will have *some* colour. Being coloured is part of the structure of *physical* bodies. Similarly, being a substance, or standing in causal relations to other substances, is part of the *metaphysical* structure of beings. In knowing such general *metaphysical* properties of beings, we also know some necessary propositions concerning them. In this sense, categories are *a priori* determinants of empirical objects.

So much for the pre-critical view. Once the critical turn has been taken, the status of the categories as *a priori* determinants of possible experience remains unchanged; however, this status is no longer explained with reference to their ontological role, for the latter could at best be noumenal, and therefore unknowable. Their role as *a priori* determinants is now explained by the Deduction and the Schematism. Because they are implicated in the synthetic acts that produce cognitions, the categories continue to characterise the structure of the latter. And thus they are

2 *Critique of Pure Reason* (hereafter *KrV*) A832/B860-A851/B879.

also the base-level determinations of any possible objects of experience. For instance, *if* something is an object of experience, then it must stand in causal relations to other objects, for otherwise these objects could not all belong to the same, unified field of experience.

So what does this say about the role of the categories in natural science? The Leibniz-Wolffian ideal of divine knowledge, that is to say of an exhaustive determination of all the properties of every substance, is preserved in Kant as the ideal of a completed system of natural science. Moreover, the categories have been assigned the role of basic determinants of all possible cognitions, whether these are scientific or not. Science differs from everyday empirical knowledge precisely in its attempt to organise its empirical concepts in a *system.* Kant explains in the Architectonic of Pure Reason that in such a system, there is "a unity of the multiplicity of cognitions [*der Mannigfaltigkeit der Erkenntnisse*], such that the scope of the multiplicity, as well as the positions of the parts, will be determined *a priori.*"[3] Thus the pre-critical picture I just described, in which the categories will stand at the root of all scientific knowledge because they are, ontologically, the root of all being, is replaced by a limit concept: in a completed system of nature, our knowledge will be organised in such a way that the categories *will play this role.* We have, as it were, a provisional assurance that this is possible, because the categories are involved in the production of each individual cognition. But it is quite evident that, from the fact that each event does have a cause, it cannot be proven that all such causes will be unifiable in the way Kant imagines in the passage I have just quoted. The most he can hope to demonstrate is that the categories *could* serve as the root of such a system—that it is at least possible for them to serve as the base concepts of natural scientific laws, that is to say as fundamental determinants of nature, whatever else may then transpire in the course of empirical investigation. Showing that this is possible is the goal of the *Metaphysical Foundations of Natural Science.*

Kant indicated this connection himself in the *Prolegomena.* The operations of the understanding are necessary and sufficient conditions for the complete determination of empirical concepts not merely in the sense of the Transcendental Deduction and the Schematism: they are also the sources of the most general laws of nature.[4] But showing this

3 *KrV* B860.

4 *Prolegomena* (Kant 1911b), 4.306. With the exception of the first *Critique*, for which I use the standard A|B numbering, all works of Kant are cited by their

will involve more than merely showing that they are involved in all empirical cognitions. Kant must show that they are adequate to the task of constructing a unified system of nature, and this is a task that must be accomplished outside the domain of strictly transcendental philosophy, one which is therefore external to the first *Critique*. The *Metaphysical Foundations* extends beyond the largely "negative" demonstration of the *Critique*, in which the categories[5] are shown to be necessary to our experience, scientific or otherwise. In this "metaphyiscs of nature," it must be shown that the categories (and the principles derived from them) are adequate to the task of constructing a complete system of nature in which all appearances are "determinate" in Leibniz's sense.

Kant's approach is, as usual, transcendental. He argues that the core concepts of natural science, which he calls the "metaphysics of corporeal nature," can be determinate only if the categories are invoked. Furthermore, these core concepts—matter, motion, mass and force—can be given corresponding schemata which are adequate to the demands that Kant places on what he calls a "proper science." In such a science, all concepts and laws are assigned a place in a unified system, and this determination is carried out *a priori*. Kant argues in the Introduction to the *Metaphysical Foundations* that these requirements can only be met if the concepts and laws in question are mathematical. Thus his task is to schematise the core physical concepts so as to allow us to construct a hierarchical system of mathematical laws of motion. The base-level laws of this system will therefore be "pure empirical" extensions of the principles of pure understanding, in which the latter are supplemented by a mathematised concept of matter.

In order to prove that he has succeeded in this undertaking, Kant will have to show that the empirical schemata of the concept of matter that he develops in the *Metaphysical Foundations* render the basic concepts of physics both mathematical and determinate. My discussion in this chapter will therefore concentrate on these two aspects of Kant's argument. In the first section, I will analyse the notion of determination in Kant's logical writings, in order to apply these purely logical definitions to the argu-

---

Akademie-edition pagination. 4.306 means page 306 of the fourth volume of the Akademie-edition.

5 I use the term here as a shorthand for the categories and the various principles of the understanding that derive from them. The exact relation between the categories, the principles of pure understanding, and the four sections of the *Metaphysical Foundations* will be explained in section (c) of this chapter.

ment of the last section of the *Metaphysical Foundations*, the Phenomenology. In this section, Kant recapitulates the arguments of the three previous sections, arguing that the categorical determinations of the concept of matter developed there progressively tighten the scope of this concept. In effect, he checks his own demonstration by verifying that the core physical concepts now satisfy the logical demands of determinacy, which amounts to demonstrating the efficacy of the preceding sections.

In the second, longer section, I shall examine the role of the various constitutive and regulative demands which Kant appeals to in the first three sections, paying special attention to his arguments for the necessity of central forces, and to his critique of Newton's law of force composition, that is to say the parallelogram law. These aspects of Kant's theory will prove essential to the reading of Helmholtz's philosophy of science that I advance in the succeeding chapters. But my choice of these topics is also dictated by the importance that Kant himself attaches to them. In insisting that the system of concepts he is developing must be adequate to the demands of a complete system of nature, Kant requires that these concepts must be mathematical, and that they must be adequate to the task of formulating general laws of nature. To see how these two requirements interact with the determinacy requirement, we shall consider the concept of force, which corresponds to the category of cause and effect.

According to Kant, force is a mathematical concept of causation (thus fulfilling the mathematisation demand), and this concept must furthermore be suitable for formulating general laws. From the latter demand, Kant can argue that the forces involved in material systems must be related, both in their actions and in their own determinants, to the distances among the points in the system. (By "determinant," I mean that on which the change of a force depends. For instance, if a force varies with position, then it is determined by position, i. e., position is a *determinant* of that force.) For if this were not the case, then neither their effects, nor their own determinants would be empirically defined. He concludes that they must be central, for (1) central forces act only to change the distances between the points among which they hold, and (2) central forces change their intensities only as functions of these same distances. Finally, Kant claims (erroneously) that (3) if forces are central, then it follows that the parallelogram law of force composition is apodictic. Thus the concept of central forces is uniquely well-suited to satisfying the determinacy and systematicity imposed on us by the architectonic demands of reason.

In the final section, I shall address the errors in Kant's reasoning, and the relation of his theory to Helmholtz's later application of it in his memoir on the *Conservation of Energy* (*Die Erhaltung der Kraft*). According to my analysis, Kant makes repeated appeals in the *Metaphysical Foundations* to the relativity of space in order to justify his claims concerning the possible forms of force laws. His central claim is that, since space is indeterminate unless there are empirically given points within it, we can draw inferences concerning the necessary centrality of force by appealing to the *in*determinacy of spatial magnitudes. For instance, Kant claims that forces can act only along the straight line connecting two empirically given points, because when only two points are empirically given, no other spatial magnitudes are determined. But such symmetry arguments are, I claim, at odds with the theory of mathematics presented elsewhere in Kant's writings. For in the *Critique*, Kant argues that geometrical propositions are not merely analytic (and thus that space is a pure form of intuition) by appealing to our need to perform what he calls "constructions" in the course of geometric proofs. Such constructive procedures involve the manipulation of spatial forms, for instance the displacement of points and lines involved in establishing relations of congruence. Thus Kant admits that there is a sense in which spatial relations are determinate even when no "empirical points" of the sort referred to in the *Metaphysical Foundations* are at hand. Space is not a purely featureless background—from our current point of view, we would say that it has a metrical structure. But if this is true, then Kant's appeals to empirical indeterminacy in the *Metaphysical Foundations* are unjustified.

It is, I shall suggest, this tension within Kant's theory that eventually becomes explicit in Helmholtz's later development of it. On my account, the problem ceases to be a purely theoretical one for Helmholtz because he is among the first people to consider the pressure placed on the role of geometry once the nascent theory of electrodynamics makes appeals to properties of absolute space. Helmholtz's concern with the epistemological status of geometry is not just a naturalist critique of nativist theories of perception, although he obviously did take it as a stroke against such opponents. It is rather of a piece with the main line of this development, which continues through the work of Poincaré up to Einstein.

(a) Conceptual Determination

Now that we have a general sense of the context within which Kant's *Metaphysical Foundations* should be situated, we may turn to the first of our tasks, which is to extract those purely logical doctrines which Kant will employ in the Phenomenology of the *Metaphysical Foundations* to check his work. In this last section of the book, Kant examines the modal status of a disjunctive proposition, each of whose disjuncts represents a distinct motive predicate. Roughly, he considers the disjunction, "Body A is in motion and reference frame F is at rest OR Body A is at rest and reference frame F is in motion" from the point of view of the Phoronomy, the Dynamics and the Mechanics of the *Metaphysical Foundations.* In each of these sections, the concepts of matter and motion have been given a tighter categorical determination, and Kant argues that the disjunction in question moves from being "alternative" to "disjunctive," and from "disjunctive" to "distributive." In order to understand why this logical check is important to his argument, we need to specify the various logical concepts that he is appealing to, and we need to get a sense of how the notion of "logical determination" that is in play here relates to the conception of a system of natural science outlined in the Architectonic of Pure Reason. This will be undertaken in this section. Once this analysis is completed, we will turn in section (b) to the Phenomenology to reconstruct Kant's claims with regard to his own demonstrations. This analysis will then permit us to address the role of constitutive and regulative principles in the schematisation of the concept of force in section (c).

Kant gives an explanation of "logical determination" in the chapter on the transcendental ideal in the Transcendental Dialectic:

> The logical determination of a concept by reason rests on a disjunctive syllogism [*Vernunftschluss*] in which the major premise contains a logical division (the partitioning of the sphere of a general concept), the minor premise restricts this sphere to a single part, and the conclusion determines the concept by the latter [part]. (*KrV* B604/A576)

Determining a concept presupposes that we have the materials at hand to construct a syllogism of a specific sort. It has two parts: (1) a major premise in the form of a (real) disjunctive division of a general concept, and (2) a premise asserting that one of these cases or parts is singled out as applicable. Since an inference of reason (that is, a syllogism) is "the cognition of the necessity of a proposition by subsuming its condition under a given general law,"[6] it follows that any such inference will involve all three of the faculties of understanding, judgment and reason; furthermore, its conclusion "is always accompanied by the conscience of necessity, and has as a result the dignity of an apodictic proposition."[7] Let me address these points in turn.

Suppose I want to determine the concept *animal* with regard to the concept *horse.* First I divide the former into exclusive and exhaustive parts, e.g. "Animals are either motile or sessile," and take this as my major premise. My minor premise might be: "Horses are not sessile." The concept partitioned in the major premise is thereby restricted to one part, and the conclusion, "Horses are motile animals," determines the concept *animal* by that of *motility,* such that the result, *motile animal,* falls under *animal,* and is contained in *horse.* Having carried out this inference, I know both that: (1) Horses are motile animals, and (2) Horses are not sessile animals. This is to know considerably more than is expressed in the conclusion, for there are other ways I could have arrived at the proposition, "Horses are motile animals." It could have been produced directly and spontaneously by the understanding, then judged correct by the faculty of judgment. In such a case, a merely problematic judgment is rendered assertoric, but because the proposition is not the product of an inference of reason (a syllogism), it does not have the latter's apodictic "dignity." Only when the proposition results from my subsuming a cognition under a general rule do I establish apodictic relations between the concepts in question. But even establishing such relations apodictically—by, e.g. the inference, "Everything motile is an animal," "Horses are motile," "Horses are (motile) animals"—will not suffice to determine *animal* with regard to *horse.* For if we were then asked whether horses are sessile animals or not, we would not know what to say. Only disjunctive inferences secure a connection between, on the one hand, the sub-concept of *animal* (*motility*) which we predicate of the determining concept (*horse*) and, on the other, those sub-concepts (*sessile*) which we thereby reject. Such a syllogism, in short, establish a local tree-structure under *animal,* and locates *horse* uniquely on this tree.

Not only does such determination require a *disjunctive* syllogism, however. If it is to deal with real possibilities, the disjuncts of its major premise cannot be mere logical contradictories, like *motile* and *non-motile.*[8] For we do not in general know what, if anything, falls under the negation of a concept. If, in determining our concepts and representations, we are trying to estab-

6 *Jäschke Logic* (Kant 1911c) §58, 9.121.
7 *Jäschke Logic* §60, 9.122.
8 *Jäschke Logic* §106, 9.143–144.

lish logical relations among them which will permit us to reason about them synthetically, then we cannot content ourselves with disjunctions that result from purely formal operations of the understanding, which will produce merely analytic truths. The disjunction forming the major premise of our disjunctive inference will have to consist of positive sub-concepts having a "real possibility," which together exhaust a general concept (*Oberbegriff*) while excluding each other.

Kant describes the parts of a *Vernunftschluss* or syllogism in the *Jäschke Logic* §58 as follows:

> The following three essential parts make up each inference of reason.
>
> (1) a general rule, called the major premise.
> (2) the proposition which subsumes a cognition under the condition of the general rule, called the minor premise; and, finally,
> (3) the proposition which affirms or denies the predicate of the rule of the subsumed cognition—the conclusion.[9]

Three distinct intellectual activities [*Handlungen*] are involved here:

(1) Forming of a general rule;
(2) Subsuming a cognition under the condition of the latter;
(3) Inference by means of the middle term (the condition of the rule, and the concept under which the cognition of (2) is subsumed).

The first step is carried out by the understanding, for it establishes relations between concepts. The second step involves the faculty of judgment, which is responsible for asserting that a given representation (or concept) either does, or does not fall under a given concept. The last step is carried out by reason (in the narrow, logical sense), for it draws the conclusion by placing the cognition of (2) under a general, unifying rule (whether *a priori* or empirical). It is by virtue of this last step that the conclusion—which could obviously have been produced immediately by the first two faculties alone—has apodictic force. But why does Kant think that by bringing a judgment under a rule to produce the conclusion, we endow it with a different modal status? And, allowing that it does acquire this status, why is it the *conclusion* that Kant views as necessary, and not the general rule from which it was derived? Today, we would say that, from the necessary general proposition $Fx \vee \sim Fx$, and the contingent particular proposition $\sim\sim Fa$, the proposition Fa *follows* necessarily.

9 *Jäschke Logic* §58, 9.120.

But the *conclusion* is not thereby necessary, whatever the modal status of the general proposition and the derivation rules themselves.

The short answer to these questions is that in deriving a proposition such as, "Horses are animals," from the general rule, "Everything motile is an animal," and the assertoric proposition, "Horses are motile," we connect the concept *horse* to what Kant calls the "exponent" of the rule. A rule consists of a "condition" and an "assertion," and its exponent is "the relation of the condition to the assertion, namely how the latter stands under the former."[10] Thus the exponent of the rule "Everything motile is an animal" is the relational structure that embeds the extension (what Kant calls the "sphere") of the concept *motile* within that of the concept *animal*, and, conversely, the (intensional) content of *animal* within that of *motile.* When we draw the inference by subsuming the predicate of the minor premise under the condition of the rule, we use this relation to establish a connection between all of the concepts *animal*, *motile* and *horse.* In contrast, the proposition, "Horses are (motile) animals," can just as well be generated by expositing the concept *horse* directly, in which case no systematic connection among the concepts need become evident. Furthermore, no matter how often I exposit the representations of similar concepts (*cow*, *dog*, *etc.*) onto the concept *animal* (cows are animals, dogs are animals, *etc.*) nothing will licence me to derive a strictly general rule concerning such concepts, nor, in consequence, concerning the concepts and entities that fall under them.[11]

Thus what makes the conclusion of a syllogism remarkable is that we draw it by bringing a judgment into relation with a rule that unifies concepts under the structure of its exponent. That this structure may be provisional, incomplete or even ill-conceived is irrelevant to the distinction Kant has in mind: in the absence of a complete scientific system of nature, the majority of general laws we work with will be flawed in one of these ways; however, they will still have the function of unifying experience by providing connections between concepts, and thus of allowing us to draw mediate (synthetic and rational) connections between judgments. They will thus permit a partial determination, and thus a partial objectification of our experience.

Now it is in just this regard that the two principal kinds of rational inferences, namely categorical and disjunctive ones,[12] differ from each other.

10 *Jäschke Logic* §58 Anmerkung, 9.121.

11 Remembering that, for Kant, there is no essential logical difference between partial and complete concepts.

12 Hypothetical inferences are, strictly speaking, not inferences at all, for they have no middle term. *Jäschke Logic* §75, 9.129.

Both general categorical and disjunctive propositions are rules; however, the structure of their exponents differ, which difference Kant explains as follows:[13]

> In categorical judgments, the *x* which is contained under *b* is also under *a:*
>
> 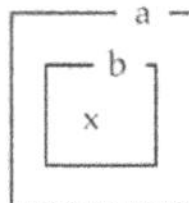
>
>
> In disjunctive ones, the *x* which is *a*, is either under *b* or *c etc.:*
>
> 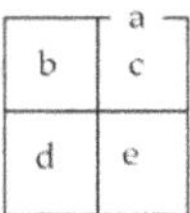
>

As we saw, the determination of a concept by reason depends on our major premise having an exponent of the second sort. For indeed it is only when the concept *a* is exhausted by the disjuncts that we can make a *universal* claim about *a* in the major premise (all *a*'s are either *b* or *c* or …). In the first case we can at best determine something about the relation of the concepts *a* and *x*, but we cannot know what lies outside the concept *b* (and under *a*), and we have no basis for asserting that *b* is the uniquely correct determination of *a* with regard to *x*. In other words, because, "All *b*'s are *a*'s," provides only an incomplete partition of the concept *a*, the sphere of *a* which is not-*b* is indeterminate. All we can say about it is that it contains whatever is *a* that is not-*b*. But, as Kant explains in the *Critique*, merely attaching the particle "not" to *b* tells us nothing about the content of *not-b:*

> If we consider all possible predicates not merely logically, but transcendentally, that is according to their content, which can be thought in them *a priori*, we find that through some of them a being is represented, whereas through others a mere non-being is represented. Logical negation, which is indicated only by the particle: Not, never properly attaches to a concept, but rather to the relation of the latter to another in the judgment, and can therefore by no means suffice to designate a concept with regard to its content.[14]

---

13 *Jäschke Logic* §30, 9.108.
14 *KrV* B602/A574.

The categorical statement, "All *b*'s are *a*'s," can be transformed into the disjunction, "All *a*'s are either *b*'s or not-*b*'s," which does indeed partition the sphere of *a*. But on Kant's view, this manipulation does not and cannot tell us anything about what actually does fall under not-*b*. In order for the latter to take on a content that was not parasitic on that of *b*, we should have to replace it with one which was extensionally equivalent, but whose intension (Kant's "content") in no way depended on that of *b*. And this is just what Kant requires of the major premises of the disjunctive syllogisms we use to determine concepts.

Nevertheless, this last requirement, as innocent as it may seem, raises a host of difficulties. For it clearly requires that, in setting up the major premise, we are somehow able to exhaust a concept by anticipating (I use the word advisedly) anything that might fall under it, and this without merely logical tricks. In other words, although the syllogism is well-formed only if its major premise has a specific logical property (it partitions its concept), we are forbidden to use logical methods to satisfy this requirement. Clearly this demand can rarely be met, and in most cases where we determine concepts, we will fail to attain this standard. To meet it strictly would require us to give pure *a priori* demonstrations of the possibility of each positive disjunct, as well as of their exhaustiveness and distinctness; and the only sciences in which this can be achieved are the mathematical ones, where we *construct* the cases in question in intuition. Nonetheless, Kant does believe that it can be met in other cases, providing we relax the requirement that the demonstrations be pure *a priori*. Indeed, as I argue in the following section, Kant's *Metaphysical Foundations*, particularly the section on phenomenology, is an explicit attempt to make good on this claim in the case of physical science.

Before turning to a detailed account of Kant's method in the *Metaphysical Foundations*, let me summarise the results of the preceding discussion. We have established that the determination of concepts, in Kant's employment of the term, is a matter of establishing conceptual hierarchies, such that each term in the hierarchy is unequivocally situated with respect both to the concepts under which it falls, and with respect to those it subsumes. The hierarchical ordering of intensions mirrors that of the extensions of the latter: if *horse* is subsumed by *animal*, so that the extension of *horse* is included in that of *animal*, then the intension of *horse* contains the intension of *animal*. These logical hierarchies are generated by the faculty of reason, which specifies the local tree-structure holding between a group of concepts by means of disjunctive syllogisms. Although it is conceivable that the same structure could be generated merely by asserting a sequence of predicative propositions, the latter procedure could only accidentally generate the kind of structure that is required for scientific cognition. For a tree with the required structure should embody apodictic relations among its concepts: propositions asserting these containment relations (intensional and extensional) must be necessary, as opposed to merely empirical or accidental.

Thus Kant puts strong constraints on this procedure. In order that the hierarchies so generated may be complete, the disjunctive proposition forming the major premise of the syllogism used to establish a local tree must be exhaustive. The concept in question must be partitioned into exclusive subsets. And the subsets must themselves be defined positively. Each term of the disjunct must have a content that is not defined in terms of the negations of the others. This means that the truth of the disjunction cannot be analytic: if the major premise expresses a necessary truth, it must be synthetic *a priori.* Of course, this will not be the case for the vast majority of such disjunctions. Much work in the empirical sciences will involve identifying the species of a genus in terms of their positive differentia, in order to establish synthetic *a posteriori* partitions of the latter. These will always be open to revision in the event that, say, a new species in the genus is discovered.

This logical model, as we shall see in the following, is crucial to understanding Kant's conception of the ultimate aim of natural science. Whereas everyday concepts are related to each other in a haphazard manner, so that their containment relations are only ever partially established, and the relations of exclusion holding among them are not rigorously known, in natural science we aim for an exhaustive and univocal hierarchy. It ought to be the case that for each concept in the natural sciences, we can say exactly where it fits in the hierarchy. Knowing where it fits will mean knowing the concepts falling under it, and thus the class of appearances to which it ultimately applies. But it will also mean knowing all the concepts under which it itself falls, and thus the class of intensions forming its definition. Were we to find ourselves in this ideal situation, we would be entitled to say that nature was fully determinate. Having subsumed a natural phenomenon under the appropriate concepts, we would be able to say, with apodictic certainty, what other propositions followed from this description, and thus we would be able to make certain statements about its relation to other possible phenomena: in the biological case, we could derive other functional and anatomical characteristics from knowledge of a restricted set of these; and in the case of mathematical physics, we could make certain statements about the evolution of a system given knowledge of its initial state. The logical relations among the concepts in the completed system would thereby stand in for the metaphysically necessary connections that are assumed in pre-critical accounts of science. For while it is true to say that nature is "deterministic" once the system of natural science is complete in this sense, the determinism in question is not a metaphysical determinism (such as that derived by Descartes from the essential properties of matter). The only necessity involved is a conceptual necessity, and while it may be true that this necessity reflects some unknown noumenal necessity holding among the transcendent causes of phenomena, such a necessity is not, and cannot be used to establish the relations holding among the elements of our knowledge. This means that in so far as one may appeal to the necessary connections embodied in the structure of the tree in doing science, such appeals can only be grounded epistemologically. Thus all synthetic *a priori* principles employed in the natural sciences will have to be given a transcendental justification, which will show why they represent what Kant calls "constitutive" or "regulative" conditions on our determinate knowledge of the natural world.

This ideal vision of a complete and univocal natural science plays a fundamental role in both Kant's and Helmholtz's theories of science. First of all, the aim of establishing such a completed system of natural science sets what Kant calls "regulative" constraints on the possible form of scientific concepts. A concept at level n in the hierarchy must be chosen in such a way as to subsume concepts one level down, at level n–1. But it must also be a candidate for subsumption under concepts at level $n+1$. That is to say, each concept, and each law expressed in terms of these must be a specification of higher-level, more general laws and concepts. Secondly, Kant holds that the highest-level laws and concepts in a natural science must be known with *a priori* certainty if the science is not to be *merely* empirical.

These demands set limits on the form of a proper (*eigentliche*) science. Since the disjuncts that establish the tree-structure must be synthetic, it follows that the base-level laws and concepts, the "root" of the scientific tree, must be synthetic *a priori.* But there are only two kinds of synthetic *a priori* principles available in the critical system: general logical principles derived from the categories, and mathematical principles resting on the pure intuitions of space and time. The first sort are, however, only applicable to experience in so far as they have been given corresponding intuitions, as is the case with the pure propositions of mathematics. We prove such propositions by applying the quantitative or "mathematical" categories to the intuitions of space and time, in which we can "construct" the required intuitions. Kant envisages an analogous procedure for the remaining, "dynamical" categories: although they are not immediately concerned with magnitudinal concepts, we can provide them with what he calls "pure empirical" constructions by introducing minimal empirical concepts. In the case of physical science, the empirical concept in question is that of *matter.*

In the next two sections, I will examine Kant's method in the *Metaphysical Foundations of Natural Science* by considering the role played by the various regulative and constitutive principles derived from the categories (from the pure operations of the understanding). The concern is above all to clarify their role in establishing a complete and determinate system of nature—the very sort of "complete" scientific theory that Helmholtz refers to when he talks about the "complete comprehensibility of nature" (*die vollständige Begreiflichkeit der Natur*) in the introduction to the *Conservation of Energy.*

(b) Phenomenology and the Determination of Motion

The fourth section of the *Metaphysical Foundations*, the Metaphysical Foundations of Phenomenology, explains how the previous three sections, namely the Phoronomy, Dynamics and Mechanics, provide a step-by-step determination of the predicate of motion with respect to the modal categories of possibility, actuality and necessity. As we saw in the preceding discussion of logical determination, the assignment of such a modal category to a judgment does not depend so much on the truth of the judgment as on our logical grounds for asserting it. "Horses are animals" could be a judgment derived directly from experience, in which case it would be true and *actual* (*wirklich*). But if it is derived by inference from a previously established conceptual hierarchy, Kant will call it *necessary.* In saying that the progressive determination of the concept of motion carried out in the first three sections of the *Metaphysical Foundations* corresponds to the transitions from possibility to actuality, and actuality to necessity, Kant is making a meta-theoretical point. Judgments of the form "The body A has speed v," where v is a magnitudinal concept, change their modal status as the conceptual framework in which they are embedded is progressively tightened. As the concept of matter is refined by successive categorical determinations, the class of possible phenomena that can be subsumed under a given motive concept is restricted. In the final stage, the Mechanics, statements concerning the motion of objects are fully determinate, and thus *necessary*, although they quite obviously retain their empirical reference. In order to see how this "downward" determination works, I will briefly summarise Kant's reasoning.

In the first section of the *Metaphysical Foundations*, the Phoronomy, Kant defines matter as "the movable in space." Here, motion is considered without regard to its possible causes (forces, which are first introduced in the Dynamics), and matter is merely something that can occupy different parts of space at different times. Thus in order to determine what we mean by a statement such as, "The body *A* has the speed *v*," we need to translate the latter onto distinct possible appearances that might realise this concept, and we must choose that unique one to which the predicate truly applies. But the single principle of phoronomy—that any motion of an object can just as well be ascribed to a reference frame as to the object—makes such a unique determination impossible. Absolute motion cannot be an object of possible experience, and thus the class of appearances falling under the concept "in motion with speed *v*" is not determinate. The predication of the speed *v* to *A* is merely *possible.*

In the Dynamics, Kant extends the concept of matter to include the notion of force, corresponding to the category of cause and effect. Introducing such a notion permits, indeed compels us to make a distinction between appearances which are indifferent alternatives from a phoronomical point of view. For when we ascribe a rotational motion to a body, we are fully justified in phoronomy in regarding this case (alternatively) as a rotational motion of a frame of reference about a body which is at rest. But since a rotational motion involves a change of a change (a change in speed), the understanding, by appealing to the "dynamical" category of cause and effect, views the two cases as distinct and mutually exclusive. Thus there is an actual question as to which representation truly realises or determines the concept of motion in question. The predication of the one or the other speed is not only possible, but *actual* (*wirklich*).

Finally, in the Mechanics, Kant extends the notions of matter and motion to comprehend a principle derived from the category of community, namely that in every mechanical interaction action and reaction are equal and opposite. If I now want to know how the predicate of motion attaches to a body in a mechanical system, I can overcome the difficulty that faced me in phoronomy. For although several representations of the system are possible, only one of these will, according to Kant, agree with this fourth mechanical law, which corresponds to Newton's third law of motion. Only in the frame of reference determined by the centre of mass of the system will action and reaction be equal.[15] Thus we are entitled to single out this representation of the system as correct—the predication of speed will be *necessary.* At first glance, this case is not distinct of the dynamical one, for in both of them, the predicate in question is univocally applicable to the phenomenon. But, Kant argues, the mechanical case differs from the dynamical one because it is based entirely on general rules: the choice of the true representation is given by a general mechanical principle, thus the determination in question is not only actual, but *necessary.*

As we saw in the previous section, in a completed conceptual hierarchy, the application of a concept to an appearance is univocal in a double sense: we know in each case whether or not the concept applies to the appearance; furthermore, we know of each specific concept that *if* it applies to the appearance, then the other specific concepts in the genus do not (for instance: if horses are motile animals, then they are not sessile animals). The connection I have drawn between the logical account of conceptual determination, the modal status of determinate (and partially determinate) judgments, and the argument of the *Metaphysical Foundations* is emphasised by Kant himself in a long footnote to the General Remark on Phenomenology:

> In logic, the *either-or* at all times designates a *disjunctive* judgment; for then, when the one is true, the other must be false. For example a body is *either* moved, *or* not moved, that is, is at rest. For here one speaks only of the relation of the cognition to the object. In phenomenology, where one is concerned with the relation to the subject in order to thereby determine the relation to the object, it is otherwise. For there the proposition, "The body is either in motion and the space at rest, or *vice versa*," is not a proposition in the *objective* sense, but only in a subjective one, and both of the judgments contained therein count as *alternative.* In the very same phenomenology, where motion is not considered merely phoronomically, but instead dynamically, the disjunctive proposition is, on the contrary, to be taken in an objective sense; i. e., I cannot, instead of the rotation of a body, assume its being at rest and an opposite motion of space. Where the movement is indeed considered *mechanically* (as when a body collides with one apparently at rest), the judgment which is—formally—a disjunctive one is to be used *distributively* with regard to the object, so that the motion must be ascribed not

15 In fact, this does not disambiguate between that frame and frames moving inertially with respect to it. Kant is tacitly relying on a further requirement, namely that in the preferred representation, the system as a whole must be at rest.

> to *either* the one *or* to the other, but each must be ascribed an equal share. This distinction of the *alternative, disjunctive,* and *distributive* determinations of a concept with regard to opposed predicates has its importance, but cannot be developed further here.[16]

To the best of my knowledge, Kant does not develop this last distinction elsewhere, although the term "distributive" is used in the *Critique of Practical Reason,* and both uses may relate to the distinction between the collective and distributive unities produced by reason and the understanding, respectively.[17] Nevertheless, Kant's concern with the exact status of the *apparently* disjunctive propositions produced by partitioning the concept of motion phoronomically, dynamically or mechanically clearly indicates his meaning. According to the first *Critique*, the categories represent necessary conditions on objective experience. The principles of the understanding, along with their pure empirical derivates in the metaphysics of nature, are accordingly conditions for the determinacy of *physical* concepts, such as those of motion and matter. Kant's aim in the *Metaphysical Foundations* is to show that the pure empirical principles of his metaphysics of nature represent the synthetic *a priori* root of the tree of natural scientific concepts. They should be revealed as necessary conditions on the determinate application of the concepts of matter and motion to phenomena. And we can verify that they do represent such conditions by checking whether the required conceptual structure obtains. The generic proposition "Body *A* has speed *v*" should resolve into an exclusive disjunction, each of whose terms is a special instance of *v*, that is to say one of its possible values. Furthermore, for any given phenomenon, it should be uniquely determinable which of the disjuncts applies. In such a case, we would be able to say that the concept of motion had been fully determined. If, as I shall argue in the sections immediately following, Kant has indeed shown that his principles are necessary and sufficient to satisfying this logical criterion, then he will have demonstrated that the pure empirical schemata of the categories do indeed represent the *a priori* core of natural science. So let us briefly unpack his reasoning by running through the three sections of the *Metaphysical Foundations* that precede the Phenomenology, in order to examine the modal status of this disjunction as successive predicates are added to the concept of matter.

---

16 *The Metaphysical Foundations of Natural Science* (hereafter, *MFNS*), 4.559–560.
17 *Cf. KrV* B610/A582; B693/A665.

i. Phoronomy

In the first *Critique*, Kant loosely divides the table of categories into its "mathematical" and "dynamical" parts.[18] This distinction carries over to the *Metaphysical Foundations*, since each of its major sections corresponds to one of the four major groups of the table of categories, and thus to the principles of the understanding correlated with these. The quantitative and qualitative categories (the first two groups) give rise to principles governing the *mathematisation* of experience, i. e. the axioms of intuition and the anticipations of perception. These principles are concerned with properties of appearances that are directly intuitable—the extension of a line segment, or the intensity of a colour. But the same is true neither of the analogies of experience, nor of their corresponding principles of natural science in the Mechanics of the *Metaphysical Foundations.* In the case of phoronomy, we are dealing only with the extensive magnitudes of space and time, for the matter of phoronomy has no properties beyond its being "movable in space." Thus these kinematic principles are quite literally "mathematical." For we can, just as in the case of geometrical demonstrations, "construct" the objects of phoronomic concepts in intuition.[19] The same is not true of the categories at work in the Mechanics, however. Neither forces, nor the relation of action and reaction can be given a directly experienceable correlate. Thus, as we shall see in our treatment of Newton's parallelogram law,[20] a central goal of the *Metaphysical Foundations* will be to schematise these higher-level concepts and laws on the phoronomic ones. For our immediate purposes, we need only grasp the following point. Each concept in the metaphysics of nature must be formulated in terms of concepts defined at a lower level. For instance, the concept of a speed must contain the concept of distance, as in the definition "a speed is a motion of a body through a distance in a given unit of time." In consequence, the construction of the concept of speed presupposes the construction of the concept of distance, which is the job of geometry. Thus it might seem that the determinacy of the higher-level

---

18 See the discussion concerning the relation of mathematical and dynamical categories in section (c).

19 Indeed, as I shall argue in Chapter 6, Helmholtz's notion of a "physical geometry," in which spatial magnitudes are considered to be indivorcible from the kinematic properties of ideal bodies, can be viewed as a species of phoronomy. On this interpretation, Helmholtz would be arguing that that the true basis of spatial concepts in physics is to be found in pure kinematics, and not in geometry.

20 See section (c).

concepts depends on that of the lower level ones in a straightforward manner.

On Kant's conception, however, the phoronomic construction of speed as "distance in a unit time" is only a first step towards the determination of the concept of motion. Since motion is, for Kant, an *intensive* magnitude,[21] phoronomic construction aims at describing those spatio-temporal appearances (extensive magnitudes, in Kant's terminology) which correspond to, or are determined by, a specific value of such a magnitude. The construction of speeds involves "producing" the spaces traversed by objects possessed of that intensive magnitude in some given interval of time. Suppose we take the proposition, "Body *A* has speed *v*" (I will ignore direction in the following discussion). Our task is to construct in pure intuition the set of those appearances that can be subsumed by the proposition. But we are immediately frustrated by the relativity of motion, which Kant expresses in the first and only principle of phoronomy: "Every motion as an object of possible experience can be arbitrarily regarded as a motion of the body in a space at rest, or as a body at rest and, on the contrary, a motion of the space in the opposite direction…."[22] Since our proposition can be given intuitive content only by describing the motion of an object relative to some given empirical space, and this space can itself be seen as in motion, we can construct the following series of apparently distinct cases:

"*A* has speed *v*" means "*A*'s speed relative to space *S* is *v*," therefore either

"*A* has speed v and *S* has speed 0 OR *A* has speed v_{An} and *S* has speed v_{Sn} OR … OR *A* has speed 0 and *S* has speed *v*"

where, $v_{An} + v_{Sn} = v$ and $0 \leq v_{An}, v_{Sn} \leq v$.

Each disjunct would seem to represent a distinct state of affairs, so that for any appearance we take to be a motion with intensive magnitude *v*, *one* of the disjuncts will subsume it. Indeed, one might think that only one of the disjuncts can represent the "true" motion of speed *v*, as would be the case if *S* were an absolute and immovable space. For then all the disjuncts except the first would be eliminated, and the con-

21 Kant's distinction between intensive and extensive magnitudes is discussed in greater detail in section (d).

22 *MFNS*, 4.487.

cept "has speed *v*" would be determined with respect to any given appearance. But even if we leave that ideal possibility aside, it still seems that the concept could be determined with regard to a specific appearance merely by *looking and seeing* which of the supposed disjuncts it falls under, that is to say, by supplying the minor premise of the disjunctive syllogism. Then the predicate would be determined with regard to that one appearance, even if not absolutely. But Kant's single Principle in the Phoronomy allows us to transform any such minor premise into one subsumable by another of the disjuncts, and therefore no minor premise can restrict the sphere of the concept to a single part. Despite its apparent form, our major premise is not a "logical division" (*logische Einteilung*) of the general concept in question, such as Kant demands in the *Jäschke Logic.*[23]

Nevertheless, the exercise is not futile. The constructions making up our disjuncts establish what Kant calls the real, as opposed to merely logical possibility of motion by telling us how the concept of an intensive magnitude "speed *v*" must be related to extensive ones in order to be given application to intuitions at all. But since each construction turns out to be equivocal, any judgment that an appearance is an empirically given instance of such a construction (as a plate is an empirically given instance of the geometric construction of a circle) can count at best as problematic or possible, i.e., it will be a judgment, but not, for Kant, a *proposition.* And since this holds for each of the disjuncts, it holds also for the concept of which they are the "exponent": the choice of whether or not to regard a given appearance as exhibiting a motion with a given intensive magnitude can be made "*durch bloße Wahl.*"[24]

ii. Dynamics

In the phoronomic case, the exponent of the concept is a merely "alternative" disjunction, with the result that the predication of the concept is "problematic" (hypothetical), so that the concept itself is merely "possible." Furthermore, in phoronomy, one uses the same quantitative categories employed in geometry, which explains why Kant feels entitled to call the phoronomic disjuncts "constructions."[25] The constructions are, in other words, strictly mathematical, but they are also empirically underter-

23 *Jäschke Logic*, 9.146 ff.

24 *MFNS*, 4.556. For the role of the exponent of a concept, see p. 30 above.

25 *Cf. Jäschke Logic*, 9.23, *KrV* B741.

mined. On the interpretation offered in the Phenomenology, the additional content of the dynamical concept of matter must change all these logical characteristics. In dynamics, the exponent of the concept should become an exclusive disjunction, so that the predication is assertoric, and the concept actual (*wirklich*). Furthermore, the content in question is no longer purely "mathematical" in nature,[26] for dynamics makes use of concepts "which cannot be given *a priori* in intuition at all, as e.g. that of cause and effect...."[27] Since in dynamics we are concerned with forces, the causes of change in motions, and these cannot be given a mathematico-intuitive correlate,[28] our dynamical representations essentially contain non-constructable concepts. The latter represent a specific empirical content, and it is because they do so that we can establish a strict disjunctive exponent for the concept of motion.

Whereas our attempt to determine the predicate of motion phoronomically foundered on the equivocity of our various cases, the phoronomic constructions of a rotational motion are no longer equivocal once we bring them under the concept of causality. Consider the proposition, "Body *A* rotates at speed *v*." Paralleling our last example, we can construct a series of representations in intuition where the motion is ascribed wholly to the body, wholly to the space, or divided among the two. But since a rotational motion involves a continuous change of motion, the understanding must assume that there is a cause of these changes, and thus that a force is at work. Kant argues (fallaciously) that since the motion of the empirical reference frame is "merely phoronomic," we can distinguish absolutely between the rotation of the body and the

26 The correspondence between the Dynamics and the categories and principles in the *Critique* is imperfect. It ought to derive its principles from the Anticipations of Perception and the qualitative categories, according to which equation it would treat of intensive magnitudes that represent the content of sensations, as opposed to their spatio-temporal, extensive form. Kant does indeed follow this plan in part: the Dynamics treats of those properties of matter that are necessary to fill space, and thus to form distinct, perceptible regions within it; furthermore, the attractive and repulsive forces that individuate matter are intensive magnitudes. But they are also, as in the passage at hand, referred to as "causes," which would mean that dynamics draws its *a priori* concept not (only) from the qualitative subdivision of the table of categories (Reality, Negation, Limitation), but rather from the third "dynamical" one. A detailed analysis of these structural analogies between the *Critique* and the *MFNS* is given in (Pollok 2001).

27 *MFNS,* 4.487.

28 This distinction is already central to Kant's discussion of motion and force in his *Wahre Schätzung der lebendigen Kräfte* (Kant 1910) §28, 1.40 ff.

rotation of the reference frame. For instance, depending on the speed at which one swings a hammer, the force exerted along the handle will vary. Thus each disjunct of our exponent can be distinguished by a different force, namely that which would be required to hold the body together *if it had* the motion specific to that case: if the space is at rest, the force will be that which offsets the centrifugal force produced by the rotation; if the body is at rest, no force will be found; and in the intermediate cases, the force will take an intermediate value. Given an apparent rotation of the body *A*, one must therefore look and see which of these possible values obtain for the force acting within the body.[29] The motion will then fall under only one of the possible disjuncts of the exponent. The determination is actual, and not merely possible, because the disjuncts are distinct and exhaustive. Furthermore, in contrast to the "mechanical" case that we shall consider immediately following, the predication cannot be asserted without our measuring an "active dynamical influence given by experience."[30] The predicate always either holds or does not hold of a given appearance. But we cannot make this determination without having recourse to the assertoric and empirical minor premise, i. e. by measuring an actual and empirical force. The *a priori* determination of the concept of dynamic motion opens a gap that can only be filled by experiential data. But the determination is nevertheless pure, in that it employs only elements derived from phoronomy and the principles of pure understanding.

iii. Mechanics

Following the pattern of our analysis so far, we know what to expect from the final step in Kant's procedure. The concept of motion should be rendered "necessary" by supplementing the concept of matter with mechanical principles drawn from the third group of categories, which themselves gives rise to the Analogies of Experience. Judgments concerning the motion of a body should thereby become apodictic judgments. But this further step may at first glance appear superfluous, for the disjunctive exponent of the concept is already exclusive in dynamics. So why does

29 For which purpose one would have to have knowledge of the attractive and repulsive forces possessed by the matter of *A*.

30 *MFNS*, 4.561.

Kant claim that it is only the third, mechanical determination that renders the concept of motion fully determinate?

The essential difference between the two cases lies in the need for a supplementary empirical determination in the dynamic case. Recall the inferential model of conceptual determination outlined in section (a) of this chapter: a concept is determined with regard to an object when the latter is subsumed under one of the exclusive disjuncts composing the exponent of the concept. In the phoronomic case, no such subsumption was possible—or, better, all were equally well possible, and thus none was objectively valid. In the dynamical case, the exponent was indeed a strict disjunction, but in order to know which of the disjuncts applied, one had to make an empirical check, an observation. In the mechanical case, no such check should be required. It should follow from the mere description of the system which of the possible disjuncts is correct. In this sense, the predication will be necessary, and the judgment apodictic. This is the sort of determination we require of the root concepts of natural science if they are to provide it with a stock of "general laws of nature."

Just as in dynamics we formed our disjuncts by subsuming phoronomical constructions under the dynamical category of cause and effect, so in mechanics we start with cases that already involve dynamic interaction between systems of bodies. As Michael Friedman has argued, Kant effectively transforms Newton's procedure in the *Principia* for determining the centre of mass of the solar system into a procedure for determining a rest-frame for a given system of bodies.[31] Given a phenomenal description of the motions of a system (which can be given from any arbitrary frame of reference), we apply principles from all three sciences to arrive at a description of their relative motions, and of the forces acting among them. But these descriptions are compatible with an infinite number of cases, each of which assumes a different rest frame (and each of which moves inertially with respect to the others). As in the phoronomical case, no assertoric proposition can distinguish among these cases, for each can be transformed into another by applying the single principle of phoronomy.[32] Nevertheless, Kant argues, the law of action and reaction[33]

31 (Friedman 1986)

32 "Every motion as an object of possible experience can be arbitrarily regarded as a motion of the body in a space at rest, or as a body at rest and, on the contrary, a motion of the space in the opposite direction…." *MFNS*, 4.487.

allows us to make the determination on the basis of a mechanical law alone, and this application of the principle takes on the role of our minor premise. The various disjuncts collapse on the single case which is compatible with the demand that action and reaction be conserved. Not only is the conclusion of the disjuntive inference apodictic, but so are both of the premises used to establish it, since both are derived from general laws. Unlike the dynamical case, we have no need of an assertoric minor premise to determine our concept in each empirical case. Therefore the mechanical determination of the concept is *necessarily* determinate for every appearance of an object in motion.[34]

(c) Regulative and Constitutive Principles

Looking back on the previous sections of the *Metaphysical Foundations* from the standpoint of the Phenomenology, we see that each of the previous sections—Phoronomy, Dynamics and Mechanics—changes the modal status of judgments concerning the motion of objects by extending the intension of the concept of matter. Matter is, in succession: something that is movable in space, something that fills space and is thereby a possible percept, and, finally, something that can change the motion

33 Along with, as mentioned above in footnote 15, the tacit requirement that each isolated system be regarded as being at rest.

34 This account of the matter obviously raises a number of problems, most notably concerning our knowledge of the masses comprising the system. For instance, if knowledge of the masses can be counted among the properties counted within our description of the system, why doesn't the same hold for the forces acting within it? By what right can mass be regarded as a concept that can, in some sense, be constructed *a priori*, whereas force cannot? Kant's answer to this difficulty is the traditional one: he defines mass as the "quantity of matter," admitting that the latter can only be determined by comparing the momentum of a body to those of other bodies moving at the same speed. But he insists that this definition is not circular, because we are dealing in the one case with "the explanation of a concept, and in the other with its application to experience" (*MFNS*, 4.540). A charitable reconstruction of Kant's position would have him claiming that, once all the principles of his metaphysics of nature are in place, both the concepts of mass and motion are determinate, whereas prior to that stage (i.e. in dynamics) the set of "implicit definitions" involved in the principles does not suffice to give any physical concept an unambiguous application. This reading would of course make Kant's claim concerning the actuality or reality of motion in the dynamical case questionable, but perhaps no more so than Kant's attempt to introduce the notion of force prior to that of mass did so in the first place.

of other objects. The original, phoronomical intension is extended by adding in the following concepts: in dynamics, the concept of attractive and repulsive forces; in mechanics, the concepts of mass ("quantity of matter") and the concept action/reaction. Kant's central claim is that the base-level, phoronomic data of experience are indeterminate when described by means of purely mathematical principles, and that they can be determinately described only once principles derived from the qualitative and dynamical categories are conjoined to these. In accordance with the arguments of the *Critique*, these principles are rooted in the very same operations of the understanding that we employ when synthesising experience by means of concepts, whether or not the concepts in question are scientific. The natural scientific hierarchical construction of concepts thus employs the same resources as does everyday cognition; however, it does so under much stronger methodological constraints. Thus it is able to produce a system of concepts that realise one of the ideals of reason—that of a unified, complete and determinate system of nature—far more rigorously than could ever be the case if the concepts were developed merely inductively.

In aiming to provide a complete system of nature, natural science verges towards a state of knowledge in which all concepts and appearances are strictly determined. Were our knowledge to take this form, it would resemble the divine knowledge supposed by Leibniz. But on Kant's theory, this unattainable state has no noumenal correlate: there is no ontological guarantee that the world has an internal structure that will allow for a description of this form. The principles that govern the scientific project—that one must render concepts maximally determinate, that one must describe appearances in terms of a hierarchy of laws—are thus *regulative*, not metaphysical. They are not, as Kant puts it, strictly speaking true. By contrast, the basic principles in the hierarchy, for instance mathematical ones, are true and certain. Thus the core principles of Kant's "metaphysics of corporeal nature" are determinations of these same constitutive principles *under regulative constraints.* In this section, we shall examine how regulative and constitutive principles interact in the *Metaphysical Foundations.*

According to the first *Critique*, both regulative and constitutive principles derive from the categories, or, more strictly, from pure logical operations of the understanding. The categories form the building blocks of logical reasoning not because they correspond to basic metaphysical properties (which was Aristotle's view), but simply because they are the root concepts employed by the understanding to subsume intuitions. But

Kant must pay a price in order to shift the categories from the ontological to the cognitive domain. The data of experience must be dissociated from the categories if his position is not to collapse into mere metaphysics. For if the concept "object" entailed categorical determinations analytically, then these determinations would be nothing other than essential properties of existing things, which was Aristotle's view of the matter. Conversely, if one denies this entailment, in order to restrict the categories to the cognitive side, then they risk becoming merely accidental properties, and there is no obvious impediment to our encountering objects that do not conform to them. Precisely because they prescribe only how we *think*, there is no guarantee that they determine how things actually *are*. Thus, in a second step, Kant argues that all data which are not categorically determined simply fall outside the manifold of our conscious experience, so that they can never become objects *of experience*.

The delicate business of establishing this synthetic *a priori* link between the categories and the manifold of appearances is carried out in the Transcendental Deduction, which we shall not analyse here. What matters to us is the result of the Deduction, and its further refinement in the Schematism. Kant argues there that, although the categories and the manifold of intuitions are heterogeneous, the Deduction has shown us that the temporal sequence of appearances is *unified* through categorical subsumption. Conversely, each category has a "transcendental schema." These schemata are operations of the productive imagination, which concatenates appearances according to their temporal organisation. The categories must be projected onto the structure of pure intuition if they are to have an empirical extension. When these schemata are extended to include spatial and material aspects of possible intuitions, they give rise to what Kant calls the Principles of Pure Understanding: (1) the Axioms of Intuition (principles concerning the magnitudinal structure of space and time), (2) the Anticipations of Perception (which concern the magnitudinal structure of possible sensations), (3) the Analogies of Experience (basic laws concerning the substantial and causal structure of reality), and (4) the Postulates of Empirical Thought (modal principles concerning possibility, actuality, and necessity).

Kant terms the axioms of intuition and the anticipations of perception "mathematical" principles, for they reflect the possibility of applying mathematics to experience. The analogies of experience are "dynamical," because they concern the substantial and causal nature of possible objects. The mathematical principles are *constitutive* of experience, in that the truths they express are strictly and necessarily true of every appearance.

The dynamical principles, on the other hand, express *rules* for organising experience, for instance, that one must correlate individual appearances by means of necessary connections. Thus the latter are regulative and not constitutive. Nevertheless, both the mathematical and dynamical principles reflect the conditions under which experience can be unified in consciousness, and as such they also express "general laws of nature."[35] The critical epistemology thus contains the germ of natural science in these principles, which Kant extends in the *Metaphysical Foundations of Natural Science* to produce a "pure empirical" metaphysics of nature.

As is so often the case in Kant, the distinction between regulative and constitutive principles drawn *within* the Principles of Pure Understanding is also applied on a larger scale. Kant calls the "transcendental ideas" regulative principles in order to distinguish them from other "constitutive" principles "from which, strictly speaking, the truth of the general rule … follows."[36] When one speaks of a regulative demand today, one generally means it in this wider sense. Like the principles of the understanding, these transcendental, or "cosmological" ideas also derive from the categories, but they result from the operation of reason, as opposed to the understanding. For instance, the faculty of reason produces the idea of an infinite causal chain, or the idea of the totality of existence, by recursively invoking principles of the understanding that are in themselves unproblematic: from "Each event has a cause," reason derives the undecidable proposition "Reality consists of an infinity of causes and effects."

In their illegitimate, non-regulative employment, such transcendental ideas correspond to rules which are not strictly speaking true.[37] Treating

35 *Prolegomena* §23, 4.306.

36 *KrV* B675. It is not clear whether the dichotomy within the Principles of Pure Understanding, according to which constitutive/regulative, mathematical/dynamical, is intended to coincide with that drawn between the transcendental ideas and the (unidentified) principles "from which, strictly speaking, the truth of the general rule … follows." To settle that, we would have to decide between two interpretations: (1) Kant takes the proposition that "each event has a cause" to be true "strictly speaking" and thus to be constitutive (in which case there would be two separate meanings of constitutive); or, (2) he holds that only the mathematical principles are strictly speaking true (in which case there would be only one sense of constitutive). Fortunately, we do not have to settle this matter for the purposes of the present discussion, and I will simply assume that interpretation (1) is valid, i. e. that there are two related, but distinct senses of constitutive and regulative at work here.

37 *Cf.* B675.

them as if they were true leads us into the various antinomies of reason, for instance into asserting both that the world must have a first cause, and that there can be no first cause. Taken regulatively, however, they perform a valuable service, for they direct the understanding to systematise its rules hierarchically, and thus to construct a unified totality of natural laws. Such a system of laws remains an ideal we aim at, for it cannot actually be completed. But if it were completed, it would amount to a complete *determination* of nature, for every phenomenon and every concept would be regarded as a determinate instance of a higher-level concept. Regulative principles can thus be seen from two points of view: as illegitimate statements concerning the totality of the natural world, or as methodological, meta-theoretical principles concerning the organisation of theories. Only in the latter sense can they be taken to be valid rules for thought.

Since our purpose here is to understand Kant's philosophy of *science*, I will not pursue these aspects of his epistemology further. We need to retain only two essential points: (1) the distinction between *a priori* principles that are constitutive (because they express an essential cognitive link between the categories and the structure of intuition) and those which are regulative (because they mandate the hierarchical organisation of empirical concepts and laws). And, more straightforwardly, (2) the claim that because both sorts of principles derive from conditions on our thought about the natural world, they also engender our most fundamental principles of natural science. In the *Metaphysical Foundations of Natural Science* Kant makes good on the promise of (2) by extending the principles of (1) to cover the domain of physical science. In order to achieve this, he needs to supplement the principles of pure understanding with an empirical content. For the principles enumerated in the *Critique* are *a priori* schemata of the categories, and they do not involve reference to anything beyond the *structure* of possible experience. Thus they form what Kant calls the transcendental part of the metaphysics of nature.[38] Once one specifies the species of natural object with which one is concerned, and thus the species of empirical concept, one gets either pure empirical physics or pure empirical psychology. Kant, and we, shall be concerned only with the former. Here, the empirical concept in question is that of *matter.*

Because the transcendental metaphysics of nature is conditioned by both regulative and constitutive principles, the concepts involved in the pure physics that Kant develops in the *Metaphysical Foundations* are sim-

38 *MFNS,* 4.469–470.

ilarly constrained. The constitutive, mathematical part of the task consists in providing empirical schemata to a series of material concepts. As we have seen, in the four main sections of the work (the Phoronomy, Dynamics, Mechanics, and Phenomenology) the concept of matter is successively "determined" with respect to the table of categories. This four-fold division cleaves closely to the analysis of the Principles of Pure Understanding. Matter is successively defined as follows:

(1) in phoronomy, as the movable in space (corresponding to the axioms of intuition, i. e., the theory of extensive magnitudes),
(2) in dynamics, as something which fills a space and resists motion (corresponding to the anticipations of perception, i. e., the theory of sensations as intensive magnitudes),
(3) in mechanics, as something that causes motion in another substance (corresponding to the analogies of experience, i. e., the "dynamical" principles deriving from causality and community), and
(4) in phenomenology, as an object of possible experience (corresponding to the postulates of empirical thought, i. e., the modal determination of empirical propositions).

This entire enterprise takes place under additional systemic constraints, which Kant outlines in the introduction to the *Metaphysical Foundations.* For instance, the metaphysics of nature that results should not only extend the principles of the understanding, but it should also produce a *unified system of nature.* It is, in other words, directed by regulative considerations deriving from the transcendental or cosmological ideas.

In the previous section, we did not consider the role of these higher-level, regulative principles. We saw that the aim of the Phenomenology is to show that once dynamical and mechanical determinants have been added to the concept of matter, the indeterminacy at the phoronomic level is eliminated, so that the modal status of physical propositions shifts from problematic to assertoric, and from assertoric to apodictic. These supplementary determinants derive from the principles of the understanding; however, in saying that we have not yet specified the constraints on their possible form. Thus our task in the next section will be to examine this aspect of Kant's arguments in greater detail. My suggestion is that at each level of the hierarchy, the form of physical principles is constrained not only by the lower-level concepts (those which will be subsumed by the new principle), but also by systemic demands deriving from the need to unify and systematise, i. e. from the regulative demands placed on the final structure of the hierarchy. Take for instance the case of phoronomy:

here the lowest-level concepts do not belong exclusively to the metaphysics of corporeal nature at all, for they are purely mathematical concepts derived by applying the quantitative categories to the pure forms of intuition. As we shall see in greater detail in a moment, phoronomy defines laws for the composition of motions (changes in distance) by representing motions as line-segments, that is to say as the distances that a body would traverse in a unit time. By means of these laws, one specifies the additive properties of speeds by "constructing" them in intuition. At the next level of the hierarchy, one defines laws for the composition of forces (the causes of changes in motion) by representing these as instantaneous motions. But not every kind of cause is admissible in physical theory, according to Kant. There is an interaction between what I shall call "upward" and "downward" determinacy requirements, which Kant (and later Helmholtz) takes to exclude all non-central forces. In order to explicate this interaction, we may take as our stalking horse the concept of force. This concept is indeed the very one that Kant himself chooses in the *Critique* to illustrate what he means by the regulative application of a cosmological idea; furthermore, in choosing it as our example, we will connect our discussion of Kant directly to the arguments of Helmholtz that we shall consider in the next chapter.

In the section On the Regulative Use of the Ideas of Pure Reason, Kant defines a force as "the causality of a substance,"[39] pointing out that this definition puts no restriction on the number of forces that we might find in nature. Its extension is as wide as the number of sets of appearances that exhibit a regularity, and which we therefore subsume under the categorical relation of cause and effect. But the regulative idea that there is an absolute and complete dependence among changes in appearance (to paraphrase slightly the fourth of Kant's cosmological ideas) leads us to try to organise these various forces as species of more basic forces, and ultimately of what Kant calls a "*Grundkraft.*" This is the "upward" determinacy requirement: not only must the concept of a specific (e.g. chemical) force determine the phenomena it subsumes, but it should also be seen as the determination of a more basic force.[40]

39 *KrV* B677.

40 In the case of Helmholtz's arguments, this principle is invoked to justify the claim that all forces observed in nature must be seen as determinations of a set of basic forces that characterise the various species of matter. It also finds specific mathematical employment in the argument that force intensity must depend on position.

A body of knowledge that satisfies the regulative demand to systematise need not qualify as a science in the strict sense. Biological taxonomies, for instance, organise concepts in determinate hierarchies, but according to Kant, they lack a *constitutive* core. And without such a core, they can never establish apodictic relations among their principles. In general, natural change (the subject of physics, in Aristotle's wide sense of the term) can only be described by strict apodictic laws if the changes are quantifiable. Thus Kant argues in the introduction to the *Metaphysical Foundations* that a natural science is "proper science" (*eigentliche Wissenschaft*) only to the extent that it is mathematical. The concepts of a proper science must accordingly be given a precise extension by schematising them on the magnitudinal structure of intuition. This schematisation should ensure that any magnitude involved in a theory should satisfy additive laws, in much the sense that we speak today of magnitudes satisfying measurement axioms. But even in mathematical physics, this task is not easily accomplished. Key physical concepts, above all that of force, are not among the strictly constitutive principles of the understanding. The concept of force belongs to those which are "dynamical" and "regulative" in that they enjoin us to correlate experiences by means of necessary connections. Whereas properly mathematical concepts are distinguished by our ability to "construct" their corresponding intuitions *a priori.* They satisfy additive laws by virtue of their extensive structure alone.

This demand can in fact be met in phoronomy, that part of the pure metaphysics of corporeal nature that is concerned with purely kinematic properties of matter. Constructing the motion of an idealised material point differs from a geometric construction in only two respects: we suppose an absolutely general "something" which is moved, and we imagine the motion taking place in a single interval of time. This empirical extension of the axioms of intuition can indeed be carried out with mathematical precision. But it is not immediately evident how the other principles of the understanding (the "general laws of nature" referred to in the *Prolegomena*) are to be mathematised in this fashion, for the notion of a cause is not inherently mathematical, as Kant insists. Thus a major goal of the *Metaphysical Foundations* is to explain how the pure empirical schemata of the so-called "dynamical" categories can be provided. This is achieved, as we shall see in greater detail in a moment, by schematising these categories on kinematic appearances, whose concepts are themselves schematised on geometric intuitions. This corresponds to the "downward" determination of these concepts: they must be given an extension composed of possible intuitions, each of which is also fully determinate.

Thus on Kant's analysis, the concept of motive force is squeezed between the upward and downward, that is to say the regulative and constitutive determinacy requirements. Forces, as causes of change, must be related determinately to lower-level kinematic concepts, such as the speed and the path of a material particle. At the same time, the forces thus defined are subject to regulative pressure from above. Because the science of nature must be a unified system of laws, all forces must be seen as special cases—determinations—of higher-level forces. Both Kant and Helmholtz believe that one can draw specific consequences concerning the kinds of forces that are possible in nature from these dual requirements. Because the ultimate referents of the motive concepts must be determinate spatial magnitudes, the motions that are caused by forces must all be *relative* motions. For every motion of a particle, in order to be determinate, must be relative to some other, empirically given particle. And because the differentia of high-level, basic forces (*Grundkräfte*) must also be properties of the system in which they act, these forces can only change as functions of the same, determinate spatial magnitudes.

As a result of these considerations, Kant believes that the centrality of the basic forces is transcendentally required. In the following chapter, we shall see that Helmholtz follows Kant closely on this score. The basic premise here is what I shall call the principle of positional determinacy: whatever spatial magnitudes are involved in characterising a system, they must be positionally determinate, in the sense that they are delimited by material points. Kant's reasoning is clearest in the one case where it is also valid, namely that of a two-point system: here motion can only take place along a single dimension, and the same holds true *a fortiori* of acceleration. Thus the force can only be "constructed" as acting along the line connecting the points.[41] That is the downward determinacy requirement. Furthermore, changes in the magnitude of the force can only be a function of the distance separating the points (the upward requirement).[42] Thus the force must be central. But in order to generalise this argument to complex systems, one must assume that it can be applied to each pair of points in the system without regard to the others. That would mean showing that the motions of the points in a complex system *can only be described* with reference to the spatial magnitudes determined by the various pairs of points making up the system. Such a claim can indeed

41 *MFNS*, 4.498–499.
42 *MFNS*, 4.518–519.

be defended for systems with up to four points (in three dimensions). But it is evidently not true in general.

The mathematical construction of motions and forces that Kant undertakes in the *Metaphysical Foundations* is flawed in just this last respect: Kant thinks that the centrality of force is logically entailed by his requirement of positional determinacy in part because he draws too general conclusions from the relativity of motion. Before examining these arguments in greater detail, however, I should emphasise that my analysis is skewed towards that aspect of the *Metaphysical Foundations* which will concern us in subsequent chapters, namely the link between central forces and two basic principles, namely the principle of positional determinacy, and what Olivier Darrigol has called the principle of decomposition.[43] According to this latter principle, all actions present in nature can be reduced to the (geometric) sum of the actions between pairs of mass-points, and thus the same holds of the forces causing them. The principle will be important in our discussion of Helmholtz, because Helmholtz comes to see it as a basic condition on the determinacy of physical theory. His reasoning is akin to Kant's: if forces could not be so decomposed, then they would have no determinate relation to the intuitable properties of physical systems, and the system of nature would threaten to become *unbegreiflich*, or incomprehensible. Although it is true that Kant has no direct counterpart to this principle, he is very much concerned with the parallelogram law of force addition, which as we shall see in the following, is tightly connected to the principle of decomposition. The challenge for Kant is to show that this law for the addition of forces is not merely empirical, but is in fact one of the *pure* empirical principles of the metaphysics of nature. Thus he tries to show that Newton's parallelogram law of force composition is apodictic. To imagine a composite force is to imagine, by definition, (at least) two point-sources acting on a third. The law of equal action and reaction, which Kant derives from the principles of the understanding, requires that we interpret this interaction by means of a centre of mass construction. That the composite force equals the geometric sum (for us, the vector sum) of the component forces is then a necessary consequence of the method of construction, and not an inductive generalisation. In consequence, the mathematical schematisation of the concept of causality will engender a principle of pure science, for the additive properties of forces will have been shown to follow from transcendental requirements.

43 (Darrigol 1994), p. 217.

(d) The Parallelogram Law

In order to understand the central importance of the parallelogram law in Kant's project, one must keep in mind his aim of providing an apodictic core to physical science. The central difficulty is that the dynamical categories—force among them—are not inherently quantitative. Providing *a priori* definitions of the additive relations between such dynamical concepts is, as a result, of critical importance. For without such definitions, the apodictic mathematical core of physics will not have been secured, and it will therefore fail to meet the requirements on "proper science" that Kant laid down in the introduction to the *Metaphysical Foundations.* In the following, I will briefly describe Kant's analysis of the parallelogram law through the four main sections of the book, contrasting his approach to that of Newton. Although my treatment is selective, I follow Kant's lead, for he begins his book by calling Newton's proof into question, and he returns to it in conclusion in order to illustrate the importance of his mechanical laws to the metaphysics of nature.

This criticism of Newton is first levelled in the Phoronomy, where Kant is concerned with the mathematical construction of what we would call kinematic concepts. The concept of matter to be constructed is that of a movable point in space, and Kant sets himself the task of defining the additive relations that hold between motions. Since, for Kant, all mathematics derives its synthetic *a priori* necessity from the structure of space and time, the two essential properties of phoronomic motion, namely speed and direction, must be schematised in terms of spatio-temporal magnitudes. He approaches the problem using the theory of magnitudes developed in the *Critique* in the Axioms of Intuition and the Anticipations of Perception, where he explains that all appearances are simultaneously *intensive* and *extensive* magnitudes. Because they are spatio-temporal, they have a magnitudinal structure deriving from the pure intuitions: they are *extended* in space and in time. And because they have a specific sensory content (a colour, a degree of hardness, etc.) they also have a particular *intensity.* These extensive and intensive magnitudes differ with regard to their additive properties. Because the representation of an extensive magnitude entails the representation of its parts, it is an analytic truth that the whole contains its parts. Conversely, the addition of the parts produces the whole with synthetic *a priori* necessity. Thus an additive proposition whose terms refer to extensive magnitudes is a synthetic *a priori* proposition. The additivity of intensive magnitudes, by contrast, requires a further specification of the addition operation, for an intensive magnitude does not literally contain lesser intensive magnitudes as its parts. For instance, colours are intensive magnitudes, in that they can be ordered in a sequence of intensity. But in representing two shades of a given colour, we do not thereby produce the colour which is their sum. That two colours add to form a third is not (yet) a synthetic *a priori* truth. Once an additive procedure has been defined, however, it may be one.[44]

In the Phoronomy Kant characterises speed and direction as *intensive* magnitudes. He does so not because he considers them to be sensations like colour, but simply because they do not literally contain lesser speeds and directions as their proper parts. In order to mathematise these concepts adequately, we need to provide them with definitions that secure determinate additivity relations. And here we are aided by a particularity of the phoronomic magnitudes. In contrast to sensations, they have an implicit connection to the extensive magnitudes of space and time: a speed can be represented in intuition by means of the distance that a material point covers in a unit of time; different directions can be represented by means of distinct paths. But these extensive constructions of the intensive concept of speed also reveal a difficulty. To say that speeds and directions can be added is to say that the same material point, at the same time, has multiple speeds and directions. And if these are to be constructed as distinct line segments, we are immediately faced with a contradiction: two line segments, if they are *distinct* extensive magnitudes, are contained neither in one another, nor in a third which would correspond to their sum. How can we meaningfully say of a single particle that its motion is the sum of two distinct motions?[45] On the one

44 If, for instance, rules are given for mixing particular quantities of paints, or quantities of spectral light, the statement that two colours mix to form a third will have a determinate value, since it may or may not be true. Whether or not it is true will depend on three factors: the physical properties of the substances mixed, the mixture rules, and the psycho-physiological make-up of the perceiving subject. Whether or not such a statement should be interpreted as *a priori* true (or false) is a subtle question, which cannot be treated here in greater detail. I discuss Helmholtz's attempts to define this additive operation and thereby the structure of the colour-space in Chapter 4. This research was, on my reading, of fundamental importance for Helmholtz's understanding of metrical relations on manifolds.

45 Kant anticipates the obvious objection that the sum of two collinear motions can be represented by laying the two line segments they determine end to end. In such a case, he argues, we would have represented two successive, as opposed to simultaneous motions. *Cf. MFNS*, 4.490.

hand, if the propositions of kinematics are to be apodictically true, they must map onto geometric truths. But the strict construction of individual motions as line segments produces geometrical magnitudes that do not embody the requisite additive characteristics. Thus we are at an impasse:

> Geometrical *construction* requires that one magnitude, or two magnitudes in their conjunction be *identical* with another, and not that they produce the third as causes, which would be a mechanical construction. Complete similarity and equality, insofar as it can be cognised in intuition, is *congruence.* All geometrical construction of complete identity rests on congruence. This congruence of two conjoined motions with a third ... can never take place if each of them is imagined in the same space, e.g. [the same] relative space.[46]

Kant avoids this dilemma by appealing to the relativity of space: we can make sense of the idea that two distinct motions are parts of a third if we imagine each of the two motions as relative to a distinct frame of reference. This "construction" of the addition of motions has far-reaching consequences. The addition of two velocities requires two reference-frames for its construction. Since these must be empirically given, a purely kinematic description of velocity addition requires at least three empirical points: the first is conceived as moving relative to the second, and this pair is then represented as moving with respect to the third.

Standing on their own, both the problem and its solution may well strike the reader as tendentious. Thus it may be useful to recall the general direction of Kant's arguments. The complete determinacy of natural science requires that all the concepts employed in physics have a determinate content. This means that we must provide pure empirical constructions of these concepts, which will specify precisely the possible intuitions to which they apply. Higher-level dynamical physical concepts, such as that of force, must therefore be tied to the lower-level mathematical concepts, such as motion and distance, whose changes are supposedly determined by the forces. But this requires that we specify the additive relations holding among these lower-level concepts. The *definientia* must be appropriately schematised if their *definiendum* is to be so as well. For instance, the additivity of force presupposes the additivity of velocities. And the latter presupposes the additivity of distances. But, Kant is arguing in the Phoronomy, these *cannot* be added in the strict sense of part-whole containment, because two distinct distances are, by definition, not coextensive. Kant's conclusion is that the addition of distances, and thus by extension, the addition of motions, instantaneous changes in mo-

46 *MFNS*, 4.493.

tions, and, finally, forces, presupposes the specification of frames of reference relative to which the motions can be constructed. Only in this case will we have successfully tied the dynamic concepts to the extensive magnitudes that ground mathematics.

Furthermore, Kant has a specific target in mind here. Both in the Phoronomy and in the third section of the *Metaphysical Foundations*, the Mechanics (our dynamics) Kant contrasts his construction to alternative proofs of the composition of motions. He evidently has Newton's proof of the parallelogram law in the *Principia* in mind. Such demonstrations, Kant complains, generate the empirical content of the concept of composite forces "mechanically," but they do not *construct* it mathematically.[47] In fact, Kant is simply mistaken in this, for Newton did indeed formulate a proof for the composition of velocities that would correspond to Kant's phoronomic construction in his "Tract of October 1666,"[48] and he is a good bit clearer than Kant in distinguishing between the phoronomic (kinematic) and mechanical (for us, dynamic) parallelogram laws. Newton's early kinematic proof uses his method of fluxions (the differential calculus) to show that every rectilinear motion is *simultaneously* a motion in any other arbitrary direction, where the magnitude of this second motion is the product of the magnitude of the first with the cosine of the angle they contain.[49] From this relation, he can derive the result that the two sides of a parallelogram correspond to the component motions of the diagonal motion. This result is assumed without mention in the *Principia* proof of force composition, which is what leads Kant to think that Newton has overlooked the need for the kinematic proof.

Newton's *Principia* proof considers a single point *P* at A (Fig. 1) subject to the action of two distinct forces N and M, which act in the directions of C and B, respectively. These forces, on their own, cause velocities represented by the line segments AC and AB, which are the paths that *P* will cover at those velocities in some given unit of time. The task is therefore to show that the line segment AD is the path that *P* will follow in an

47 *MFNS*, 4.494.

48 Printed in (Cohen and Westfall 1995), pp. 377–385. Richard Arthur provides a thorough treatment of this early parallelogram proof in (Arthur 2009 forthcoming).

49 In both cases, the motions in question are considered to be instantaneous. Had Kant known of this text, he might still have objected that Newton's kinematic proof, because it deals with the motion of a single point, must still make reference to the geometrical properties of an empty, and thus indeterminate background space.

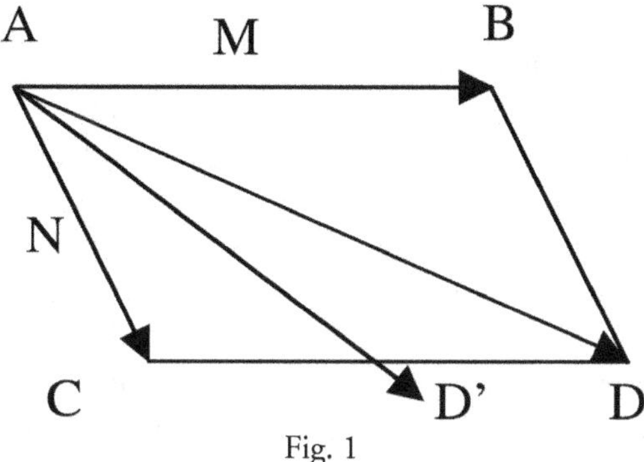

Fig. 1

equal unit of time if the two forces are applied simultaneously. Newton argues first, that *P* will take the same time to reach any arbitrary point on BD whatever the force N along AC. This follows from his Law 2, which asserts that the actions of the forces are only "along the direction in which they are impressed." And the same holds true of the time taken to reach CD. If the two forces are then impressed simultaneously, *P* will have to lie both on BD and on CD at the end of the unit of time, thus it will be at D, meaning that it will have followed the path AD. This will be, in turn, proportional to the magnitude and direction of the velocity of A caused by the joint action of the forces. Finally, on the assumption that all these velocities are interpreted as instantaneous accelerations, AD will also be proportional to the combined force causing the combined acceleration. Thus the action of the two forces acting together will be the geometric sum of their individual actions.

Now, as I mentioned above, Kant criticises Newton for giving a *dynamic* proof when he should have given a *kinematic* one: before one can show that the causes of velocities add geometrically, one needs to show that the velocities themselves do so. And indeed Newton's proof makes an implicit appeal to such a result, namely at the point where Newton maintains that "since force N acts along the line AC parallel to BD, this force, by Law 2, will make no change at all in the velocity toward the line BD which is generated by the other force."[50] Law 2 tells us that (1) a force acts only in the straight line in which it is impressed, and (2) when a force causes a motion which is "oblique" to the motion that a body is already undergoing, this new motion is "combined obliquely and compounded with it according to the directions of

50 Newton, *Principia*, Law 2, Collary 1. (Newton 1968), p. 21.

both motions."[51] From the application made in the proof, we can infer that by "compounding according to the directions of both motions," Newton means adding the velocities according to the kinematic parallelogram law, i. e. adding the instantaneous changes in distance geometrically. He takes it that when velocities are so added, it follows that whatever the velocity of *P* in the direction AC, *P* will reach BD in the same length of time. But this assumes that we know that the addition of the velocities represented by AB and AC *do in fact produce* the velocity represented by AD. Why shouldn't it be the case that they produce AD'?

In the "Tract of 1666," Newton proves that AD is the composite motion resulting from AC and AB by differentiation. Assume that *P* is moving towards AD. Then we can calculate the instantaneous rate of change in the directions of C and B, respectively, and thereby prove that the first motion is, at the instant, identical to the sum of the other two. But there are evidently two quite different ways of interpreting such a demonstration: (1) The motions are taken to occur in empty space, in which case all three directions and speeds must be assumed to be determinate relative to an absolute space.[52] The demonstration thus depends only on geometrical properties of the background space. (2) The motions are all motions relative to the three empirically given points, D, C, and B, in which case the speeds describe the rate at which the pairs of points approach one another, and the directions are given by the lines connecting them. These alternatives do not affect the mathematics involved in the proof directly, but they do affect the interpretation one must make of it when applying it in the proof of *force* composition. Suppose one takes the first interpretation: Both directions and distances are defined absolutely, and so the force acting on a point can be characterised by the direction and magnitude of the motion it causes in a mass of a given magnitude. These values are independent of the motive state of the particle at the moment the force is impressed, and so the force is a metaphysically independent entity whose two characteristics are themselves defined in terms of the properties of absolute space. On the second interpretation, by contrast, the direction and magnitude of a force can only be defined in terms of the distances

51 Newton, *Principia*, Law 2. (Newton 1968), pp. 19–20.

52 This means only that it is assumed that there is a fact of the matter concerning their speeds and directions. Newton can allow that the motions are all being constructed in some arbitrarily moving frame, but it must be possible to speak of directions and distances within this frame without reference to other empirical points.

and directions of the empirical points involved in the construction. It is, by definition, a relation between (at the very least) *pairs* of points.

For Newton just as well as for Kant, the dynamic parallelogram law presupposes a proof of the kinematic one. The latter can be provided either by differentiation within a single frame, so that one follows Newton in allowing that the same point can have multiple (instantaneous) motions at the same time. Or one denies this possibility, arguing with Kant that the motions must be relative to distinct frames, each of which must be determined by at least one empirical point. Since Kant doesn't know of Newton's kinematic proof, he falsely thinks they disagree on the very need for one. But the crux of the matter concerns the empirical determinacy of the motions being added. Newton does not *require* that motions be directed towards empirical points (though of course his demonstration is compatible with this interpretation), whereas Kant does. In so doing, Kant lays the ground for the claim that forces cannot have absolute directions, a claim that Helmholtz later uses to argue against field-theories in electrodynamics.

This phoronomic analysis is intended to feed directly into the definition of force that Kant provides in the next section of the *Metaphysical Foundations*, the Dynamics. Here, a force is defined as the capacity of one body to "resist the approach," or "to cause others to move away from it."[53] Finally, in the Mechanics, force is characterised as the capacity of a body to change the motion of another through its own motion.[54] Kant introduces a law of equal action and reaction, which in turn permits us to transform the kinematic definition of the composition of motions into a dynamical one (in Kant's terminology, a phoronomic into a mechanical one). According to this *a priori* "Law of Mechanics," the dynamical interaction of two bodies entails the motion of *both* of these with respect to that reference frame in which momentum is conserved, namely that determined by the centre of mass of the bodies. This frame of reference, Kant explains in the Phenomenology, can effectively stand in for

53 *MFNS,* 4.498.

54 Kant's "dynamical" and "mechanical" definitions of force are not truly distinct, even though Kant maintains that they differ in that the dynamical definition "could regard the matter [i.e. the body causing the motion of the other] as at rest" *MFNS,* 4.536. In fact, the strict relativity of motion postulated in the Phoronomy makes this distinction spurious. Kant's reasons for distinguishing between his dynamical and mechanical concepts of force derive from his desire to ensure a strict correspondence between the four principles of the understanding and the four sections of the *MFNS* outlined above.

Newton's absolute space: it provides an empirically determinate space relative to which the motions of bodies are themselves fully determinate. So the successive introduction of constitutive principles under regulative constraints, which consists in each case of applying a principle of the understanding to the concept of matter, determines the concepts of force, matter, and motion completely.[55]

On reflection, however, we can see that there is an essential tension at the core of this key argument in the *Metaphysical Foundations*, which in fact can only be eliminated by insisting that all forces derive from empirically given mass-points. On the one hand, the laws for the addition of forces depend on the determinacy of motions. Conversely, Kant argues that no motion is determinate until all of his principles (all those introduced in phoronomy, dynamics and mechanics) are in place. Kant's solution to this dilemma is evident in his rejection of Newton's interpretation of the parallelogram law. For Newton, forces are independent entities, which can be meaningfully said to have directions and intensities relative to absolute space—indeed he takes their presence to indicate the difference between real and apparent motions. For Kant, by contrast, every such appeal must be disallowed. If force additivity is to be schematised on the additive properties of velocities, and thus, ultimately, on spatial magnitudes, then all of these magnitudes must be grounded in intuition. Both the actions of forces, and the changes they themselves undergo,[56] must be related to pairs of mass-points. This demand corresponds to what I called earlier the requirement of positional determinacy: whenever we speak of a force, and of the motion it causes in an empirical body, there must be at least one other mass-point *towards or away from which* the motion and the force are directed.

Indeed, it is on just these grounds that Kant argues that forces must be central. On his mechanical definition, a force obtains when one body affects the motion of another. Since in Kant's pure metaphysics of nature "we regard each of these only as a point" it follows that the motion the one body causes in the other,

55 As Michael Friedman has argued in (Friedman 1986), Kant effectively inverts the relation between force and absolute space suggested by Newton, and argues that true motions, thus the concept of absolute space, are definable only within systems conforming to Newton's third law.

56 Kant also argues, as Helmholtz will do far more forcefully, that the intensity of a force can depend only on the distance separating the pair of points, *MFNS*, 4.519–521. For both men, this is a further consequence of positional determinacy. This claim will be treated in greater detail in the next chapter.

> ... must be seen as taking place along the line connecting them. But there are only two possible motions along this straight line: one in which these points *move apart*, and one in which they *approach* each other. The force which is the cause of the first [sort of] motion is called a *repulsive force*, and that of the second is caused an *attractive force*. Thus we can conceive only of these two sorts of forces as those to which all motive forces in nature must be reduced.[57]

Positional determinacy requires us to construct the concept of force as the cause of change in the relative position of (at least) two mass points, for without reference to these, there is no observable motion in the first place. Moreover, two points determine only one spatial magnitude, so they can change their relation only as this single magnitude changes. The argument can in fact be extended to systems of several points, so long as their number does not exceed the dimensionality of the space in question by more than one. But it does not hold generally.[58] The moment we consider more complex systems, the possibility of reducing the interactions among the points to interactions of pairs is no longer apodictically given. Thus Kant's claim that forces are constructable only as central forces is, as I suggested previously, an unjustified inference from the relativity of motion and the requirement of positional determinacy—an unjustified inference that Helmholtz will also draw. This means, in turn, that Newton's parallelogram law is not apodictically true, but that it is in fact, as Newton had maintained from the beginning, a proposition resting on merely inductive grounds.

Nevertheless, so long as one restricts oneself to the simple cases Kant discusses in the *Metaphysical Foundations*, it can indeed appear that forces *must* be conceived as acting along the lines connecting pairs of masses. From here, it is a short step to the conclusion that in order to "construct" the *dynamic* parallelogram law, we must imagine at least three points: that which is subject to the two forces, and the two which determine the directions along which the forces are taken to be acting. For, according to Kant, without these supplementary points it would be meaningless to

57 *MFNS*, 4.498.

58 Suppose I know how three non-collinear points in the plane have changed their relative positions. Then, from my knowledge of the motion of some fourth point relative to two of these, I can infer how it has changed its position relative to the third. I am no longer free to regard its motion as the sum of three independent variables that depend only on its distance to the other three points, because the relations among the first three points—whether expressed in angular or linear coordinates—span the space in question.

speak of the component forces at all. Finally, we can apply the results of the Phoronomy, where Kant demanded that resultant motions be strictly "congruent" to their components. Since this demand must hold in the dynamic case as well, it follows that the two motions of the particle relative to the two points that determine the forced motion sum to form a single motion that is *identical* to these. That is to say, all three motions (the component motions, which are relative motions of the pairs of points, and the resultant motion) must be conceived as relative to a further space. This demand is satisfied just in the case of a centre of mass construction, for here the motions of a point relative to the centre is indeed the geometric sum of its motions relative to the other points in the strict sense demanded by Kant in the Phoronomy: they are one and the same motion described variously with regard to distinct frames of reference.

I should emphasise that Kant does not take this last step explicitly. Those parts of the Mechanics which concern the necessity of a centre of mass construction do not mention the parallelogram law. However, when one combines the Remarks on the Principle of Phoronomy with his later treatment of the Fourth Principle of Mechanics, the link I am arguing for is relatively clear. In the Phoronomy, the claim is that in order to add two motions, I must regard each as taking place in a distinct empirical space, in order that the total motion can be strictly congruent with its two components. In the Mechanics, Kant argues that in each mechanical (for us, dynamical) interaction, the resultant instantaneous motions must be referred to a background space anchored on the centre of mass. But if this is the case in a two-point interaction, then it must also be the case in a more complicated system: here as well, the centre of mass will be the reference frame relative to which each acceleration can be represented as equal and opposite to the accelerations of the other particles. Kant's claim, in the concluding remarks to the Mechanics, as well as in the General Remark on Phenomenology, that the centre of mass of the universe anchors such an "absolute" frame of reference, makes it clear that he is of just this opinion.

Thus on Kant's analysis, the truth of the dynamic parallelogram law follows from the very possibility of constructing accelerated motions in intuition. We tend to see things the other way round. In a conservative system where all forces are central, the action of a particle relative to the centre of mass can be arbitrarily decomposed into the geometric sum of its actions relative to the centres of mass of arbitrary (and disjunct) subsets of masses in the system. All the actions sum, and can be decomposed, geometrically. But this says only that the parallelogram law of forces, when supplemented with the assumption that forces are central, will refer the total force acting on a point to the individual forces centred on the other points. And then, since momentum conservation is assumed, both the individual and the composite forces will determine actions relative to the respective centres of mass, which actions sum as do the forces—in accordance with the parallelogram law. Kant turns this deduction on its head. He too assumes conservation. He claims that in order to "construct" a composite motion, we must assume the existence of point-masses relative to which each component is determined. The phoronomic parallelogram law then demands that the total motion be referred to a frame of reference *in which it is true* that the composite motion is the geometric sum of the components. Thus actions must sum geometrically, and forces, which are the causes of changes in the relative displacement of pairs of masses, do so as well.

Now, as I have already pointed out, Kant does not make explicit the connection between the dynamic parallelogram law and the existence of the inertial frame. Thus we would be well advised to state his position in a manner that avoids insisting on this point. The easiest way to do this is to recur to the initial problem concerning the additivity of extensive magnitudes that I developed earlier in this chapter. What was the original motive for Kant's constructive project? The natural sciences, if they are to be at least in principle completable, must aim for a state of knowledge in which every appearance is determinate. Thus the aim of mechanical physics is a sytem of laws in which every appearance (every empirically given system) can be seen as an instance of a general law. In an incomplete system, by contrast, full knowledge of the state of a physical system would fail to give us knowledge of its future motion. It is just this undertermination of physical concepts that Kant highlights in the Phoronomy: here the appearance of motion can always be reinterpreted as a state of rest combined with the motion of a reference frame. Whereas, as we saw in the section dealing with the modal status of such disjunctions, when a system of principles is adequate to the aim of determining our basic physical concepts, this essential ambiguity drops out.

Thus the result of the Kant's demonstration up until the Mechanics should be that when all the basic principles of the metaphysics of nature are in place, knowledge of the masses, the relative motions, and the forces acting within the system should be adequate to characterising its state unambiguously (even if Kant does equivocate on whether or not his attractive force is inversely proportional to the square or the cube of the distance over which it acts). The analysis of the conservation law and the centre of mass construction in the Mechanics is an illustration of this process. Here, our knowledge of mass and velocity allows us to say unambiguously what the "true" state of the system is, so long as we apply the law of equal action and reaction. The more general claim made in the Phenomenology, that the centre of mass of the universe provides a frame of reference in which all motions are determinate, represents a radical extension of this view.

This all assumes, however, that the frame of reference whose existence is postulated in the last step must exist. Furthermore, as I have already pointed out, it is not in general the case that the action of a point can always be decomposed into (conservative) independent linear actions relative to the other points in the system. And these two demands, taken together, give a clear indication of what Kant wants, but cannot have. One way of defining an inertial frame is as a frame in which every acceleration corresponds to an impressed force.[59] By demanding that every motion be represented as a motion relative to (at least) one other point, Kant is able to argue that every *change* of velocity be similarly so represented, and thus in turn that every force be so as well. Since he tacitly assumes that for any system of mass-points there will be a centre-of-mass frame in which all actions sum according to the parallelogram law, he effectively equates the truth of that law with the existence of such a frame. Now, what are the presuppositions that are essential to this conception? Aside from the requirements of relativity and positional determinacy that we have already discussed at length, Kant makes two essential assumptions:

(1) The moment that I have two points, they define a line, and thus a path, along which – and only along which – their motion is representable. This line is, one might say, that straight line in spatial intuition that the pair of points selects. Kant believes that it goes without saying that such a line is an inertial path, and that such a path stands in relations to other lines that could, in the presence of other empirical points, be selected in this manner. A set of empirical points therefore defines a set of lines which are their inertial paths – those paths that they would have to follow if we were to represent their motion *at all*, that is to say independently of considering the actions of forces. This peculiar conception explains in

59 *Cf.* (DiSalle 2008), Section 1.6. "The Emergence of the Concept of Inertial Frame." Kant assumes not only that such a frame is apodictically required, but also that the accelerations defined in such a frame all resolve onto independent linear displacements relative to the other bodies in the system.

large measure Kant's dismissal of the law of inertia, which according to him is not a law at all. For it expresses, from his point of view, a truth that derives from the mere possibility of constructing the motions of empirical points (in the phoronomic sense) in intuition. Thus, in contrast to Newton, the points in the system determine *a fortiori* their inertial motions with respect to one another. For Newton, by contrast, the fact that points will move inertially along a straight line (which is defined as an unobservable subset of absolute space) is an empirical statement, which is however unverifiable on its own.

(2) Given a system of points that exert forces on one another, the magnitudes and directions of these forces are proportional to the masses of the points and the magnitudes of their accelerations. And these accelerations, if they are represented as instantaneous changes in motion, must *also* be conceived as taking place along the lines defined by the pairs of points. For since there is no way of constructing the instantaneous motions except along these lines, the same must hold of the causes of these instantaneous motions. Now, if it is true that there always exists a frame in which actions add in accordance with the conservation of momentum, it then follows that the parallelogram law is apodictically true.[60] The hidden assumption is therefore that such a frame always does exist, whereas in fact there is no impediment to imagining motions of a system that violate this constraint, precisely because the forces acting among the points, although perfectly well representable (and although in accordance with the conservation of momentum), are nevertheless not central. In such a case, one would have to choose which of the suppositions in question was to be dropped: momentum conservation, force centrality, or the parallelogram law.

If on the other hand we follow Kant in viewing the assumptions made in (1) and (2) as valid, what we get is a theory which views both the law of inertia and Newton's third law of motion as apodictically true. And this has extremely important consequences, both for Kant's theory, and for the use of it that Helmholtz subsequently makes of it. In the first instance, it sheds light on the curious relation within Kant's philosophy between the sciences of phoronomy and geometry. As I shall explain in my concluding remarks to this chapter, Kant's distinction between these two sciences is ultimately untenable, for he appeals to the *indeterminacy* of spatial relations when articulating his single phoronomic law, although he implicitly

60 Or, conversely, that if the parallelogram law is true, then such a frame must exist; however that is not the order of priority in Kant's reasoning.

assumes, when applying this law, that certain basic geometrical relations – in particular the metrical relations holding among straight lines – are themselves determined by the possible "juxtaposition" of ideal mathematical bodies. Because he does not see an inconsistency here, he also thinks that, for instance, the mere presence of a pair of bodies serves to determine a single possible path for their relative motion, and thus makes the law of inertia superfluous. This incoherence has its counterpart, as we have seen, in Newton's system, for the latter is also obliged to regard geometric relations as in some sense determinate, although unobservable, if the law of inertia is to have a *meaning*. Thus one could say that this difficulty creates, for Newton, a semantic problem within physics; whereas for Kant it generates an epistemological problem within mathematics – the very one, I shall argue in later chapters, that Helmholtz eventually confronts.

The second consequence that emerges from Kant's theory concerns the relation between the parallelogram law and the existence of an inertial frame. The trick to Kant's approach, we may recall, was to insist that the mere constructability of force additivity entailed the truth of the parallelogram law. But such a claim founders on the simple fact that, when a system is sufficiently complex, we can perfectly well imagine a forced motion with respect to that system that does not resolve onto the geometric sum of the actions relative to the other individual points. Insisting that it must do so is to make an *empirical claim* regarding the centrality of force and the truth of the parallelogram law. But if we allow Kant this step, then what he will have demonstrated is that the requirement that forced motions be empirically determinate entails the conclusion that a single privileged inertial frame exists, namely that defined by the centre of mass. In this frame all forces will correspond to accelerations, and, furthermore, the paths constituting inertial motions will always be definable with respect to the mass-points in the system itself. In other words, the core of Newtonian physics would be shown to be regulatively necessary, whereas the specific form of attraction laws would still be a matter for empirical observation. Finally, since the empirical determination of the required frame could be secured without any appeal to an absolute space, the concept of force would be freed of any metaphysical underpinning. In particular, the suggestion that forces are causes that have a claim to existence extending beyond their epistemological role in determining our experience would be undone.

(e) Conclusion

In the following chapters, we shall see how the line of argument developed by Kant in the *Metaphysical Foundations* is extended by Helmholtz, and how the latter is thereby led to problematise the epistemological status of geometry itself. On my reading, Helmholtz began these investigations because he borrowed heavily from Kant—so heavily that he inherited the implicit problems in Kant's theory of space. His later attempts to render geometry "empirical" can therefore be seen as efforts to repair the gaps in Kant's system. Nevertheless, one should guard against thinking that this development is purely internal, that is to say that it arose from purely philosophical reflection. For the difficulties in question needn't ever have come to light, and they did so only because there were experimental grounds for examining the suppositions that led to them. Helmholtz would not have addressed the problem of geometry had he not been moved to do so by specific theoretical difficulties.

Much of the pressure on Euclidean geometry's claim to unique validity in the decades between 1880–1900 derived from advances in empirical science, that is to say from the very same theoretical difficulties in electrodynamics which moved Einstein to borrow Helmholtz's concept of a physical geometry in order to operationalise units of distance and time. It was the failure to account for electromagnetic forces without introducing concepts defined in terms of absolute position (for instance, motion relative to an ether), when combined with the fact that absolute position was empirically unobservable, which forced Helmholtz and his successors to reexamine the role of geometry in physics. The problem they confronted was this: If space is conceived as a real entity, then absolute space and absolute motion are permissible concepts, and geometry can be conceived as the science of the structure of absolute space. The law of inertia therefore makes a significant (true/false) assertion, even though "true" or absolute inertial motions are never observable. Put simply, if straight lines and distances exist absolutely, the law of inertia can be objectively true, even though it is not empirically verifiable. But if, on the other hand, one denies that the notion of absolute space is permissible, accepting instead the Kantian thesis that space is merely the formal structure of possible experience, then what geometry describes are the formal properties of possible appearances. The question is then: Which of those properties that constantly accompany external appearances are properly formal in this sense? By the time Poincaré and Einstein were considering this problem, the reasons for preferring strict relativity were no longer purely philosophical, for the Michaelson-Morely experiments had seriously damaged the plausibility of the first, absolutist interpretation.

For Helmholtz, on the other hand, the matter was to a large extent a matter of philosophical preference. He did not have a compelling experimental result in front of him that would force him to pose the problem of space. But he was aware of the first theories that pointed in that direction. Thus it would be most accurate to say that he was using an argument derived from Kant to block a theoretical avenue which he felt, on philosophical grounds, to be misguided. In order to do this, he had to accept an essential ambiguity in Kant's theory, which derived from the claim that certain properties and relations of physical systems are "determinate" the moment they are conceived or intuited, and that congruence is among these. Such properties are inherent to the structure of spatial intuition, yet they are nonetheless unobservable (and thus *empirically* indeterminate) unless some empirical intuitions are given. Kant has, in other words, two kinds of basic spatial properties: the empirically determinate ones appealed to in phoronomic constructions, and the non-empirical ones which are revealed in mathematical construction. He requires these to be distinct of one another, and yet there is no obvious sense in which phoronomic and mathematical constructions *are* distinct. Indeed, this problem was already implicit in the theory that I have described in the preceding pages. Thus before we turn to Helmholtz's development of Kant's epistemology of the sciences, we would do well to specify where it lies concealed.

If we drop Kant's technical language, we might describe the view he is arguing for as follows: space is not an object of experience, but a featureless expanse in which experience plays out. If we represent the processes underlying our experience by means of point-systems, and if we require that all the concepts used to characterise the motions of these point-systems be empirically determinate, it follows not only that these concepts must *refer* to observable properties, but also that they can *depend* only on them. Since, however, the notion of an observable property is very narrowly defined in Kant's system, the claim is a strong one: both the centrality of force, and the parallelogram law of force composition are seen to follow from it. They are regulatively required, because only on the assumption that they are true will it possible to determine the laws governing the motion of a system by its purely extensive characteristics. Kant demands that a point-system must effectively be a picture of its own future motion. That is to say, the lines connecting the particles at the instant are the vectors describing the system's future motions. This model, as we shall see in the next chapter, is carried by Helmholtz to its logical extreme.

Kant sees no alternative to this view, because he believes that if a system's motion were not solely determined by its extensive properties, we would be forced to admit transcendent causes of our experiences. Since we cannot, for regulative reasons, admit this, it follows that forces must be central. In part, this conviction rests on a simple error: central forces are not the only kind of forces that could derive from positional relations alone. But this error makes Kant's philosophical motives all the more explicit. Despite the fact that he is a dynamicist, and not a mechanist, he remains a *determinist.* Causation may well be a pure operation of the understanding, without which we could never synthesise our experience. And no system of nature can do without the notion of force. But this causation should not be occult. Its sources and its effects should be completely reflected in the extensive magnitudes of time and space.[61] Of course, by "sources" we do not mean its metaphysical, noumenal grounds, but rather the circumstances under which it is occasioned. One might, using Leibnizian language, say that the extensive properties of the system must constitute a sufficient reason for its future motion. Kant is led to demand that the properties of a system that qualify as determinants of the forces acting within it should all be properties of the system itself, that is to say of the spatial relations among its material points. This is the *epistemological determinism* to which he adheres.

This division between admissible and non-admissible determinants of forces is drawn by means of a distinction between the internal and external properties of an empirically given system of points: determinations that depend on absolute relations (absolute directions, absolute motions, absolute orientations) are disallowed *a priori*, whereas those which depend only on internal properties, such as the relative positions, or the masses of the points, are permitted. In so arguing, Kant apparently commits himself to a particular view of spatial relations, according to which (1) the internal relations among empirical points *are* determined once these points are given in intuition, but (2) relations with regard to other, purely mathematical points are not. But this interpretation is hard to square with those passages where Kant seeks to prove that space is an *a priori* intuition by appealing to *external* properties: the

61 Once again, Kant cannot plausibly meet his own demands even when it comes to the concept of mass. The point here is not to defend his arguments, but to clarify their aims.

fact that the space surrounding a body can be extended in our imagination, the fact that we can imagine space empty, but not a body without a surrounding space, the fact that mirror images are conceptually identical without being congruent—all of these are taken to show that space is a transcendental condition of appearances.[62] This entire line of argument is intended to prove that space is not a mere property of things by appealing to the fact that it grounds properties that can be meaningfully predicated of them *despite* these properties being external. Kant permits, indeed he requires the geometer to construct lines at will in order to prove propositions about the internal properties of a figure by generating further properties that they do not possess analytically. Why then should the physicist be denied the same freedom?

His reasoning, in modern terms, is the following. Certain propositions concerning the properties of space are objectively true, even though these properties are not empirically realised. The latter include its topological structure, its dimensionality, and its metric. Any empirical object that appears in space partakes in all these structural properties, and, therefore, geometrical propositions can truly be said to hold of empirical objects as well. Other spatial properties lack this characteristic. Position is only relative, and, according to the *Metaphysical Foundations*, if not the *Prolegomena*, orientation and direction are so as well. These properties are not yet determined when an individual empirical point is given. The pertinence of this distinction is clearly evident in Kant's conception of phoronomy, which is an empirical science resting on purely mathematical construction: phoronomy employs a concept of matter whose schema is virtually indistinguishable from those of geometrical concepts, when one remembers that the concept of congruence does involve the notion of displacement. The task of phoronomy is to construct the concept of composite motion in such a way that the objectively real properties of the spatial manifold, namely those revealed in geometrical constructions, can be applied to derive certain kinematic principles apodictically. The only complication encountered here does indeed derive from a certain indeterminacy of spatial relations (that is to say from their relativity), but this indeterminacy derives from the objective unreality of absolute position only.

Now, my use of the term "objective" in the above summary is admittedly blunt. After all, to the extent that these properties are being predi-

62 *Cf. Prolegomena* §13, 4.285 ff. where Kant argues from the asymmetry of "internally" congruent objects for the aprioricity and independence of spatial intuition.

cated (or not) of a form of intuition, they cannot be objective in the sense that they are predicated of objects or appearances, even if it is Kant's intent that they should (also) hold necessarily of the latter. Pure intuitions are not, after all, objects. Nevertheless, we (and Kant) need some way of distinguishing between properties which are not intrinsic to intuition (position, perhaps orientation), and those which are (topology, dimensionality, metric). It is above all the metrical properties that are of significance to us.[63] For they are what permit us to say of any two quantities synthesised in intuition that they are, or are not, congruent. This relation is important not only because, as we shall see, it is that which Helmholtz later takes as the foundation of his work on geometry. Kant himself, we may recall, insists in the Phoronomy that *if* we can schematise our physical concepts on spatial magnitudes so that that demonstrations based on congruence are possible, *then* we are permitted to draw apodictic conclusions concerning the relations between these concepts.

This is just another way of making the same point from above: metrical properties must be determinate the moment an empirical intuition is given to us; they should not, as in the case of position, depend on further empirical determinations. And Kant is indeed explicit on this point: the metrical relations between empirically given external (i. e. spatial) appearances are "produced" by the productive imagination whenever an intuition is synthesised in conscious experience under the quantitative categories.[64] This does not mean that when, for instance, two line segments lie before us, we know just by inspection whether or not they have the same length—that can obviously only be decided by measurement. However, the quantitative relation (the ratio) between the two magnitudes is implicitly determinate, in that once we have decided on a unit, the result of such a measurement is preordained.

The entire argument of the *Metaphysical Foundations* rests, as a result, on the need to distinguish between (1) spatial properties which are geometrically admissible, but empirically indeterminate, and (2) properties which are determinate only when an appearance is empirically given (or empirically constructed) in intuition. In geometry, for instance, we *can* arbitrarily divide the plane into two parts to the right and left of a

63 Kant does not problematise dimensionality, and although there is an important relations between topological and metrological properties, a proper treatment would exceed the bounds of our present concerns.

64 See (Sutherland 2005), pp. 132, 143, for the role of the productive imagination in the determination of magnitudes.

given line. There is therefore no *mathematical* inconsistency in imagining that, for instance, all objects in nature are subject to a universal force in a particular direction. But by appealing to epistemological considerations, such as the principle of sufficient reason, we might argue from symmetry requirements that this case cannot in fact occur. Thus it would appear that we can also single out the class of properties (2), which are a subset of those properties considered in geometry. By insisting that only these empirically determined properties are admissible in physics, Kant establishes the claim that forces must be central: forces determine, and are determined by, only the properties of class (2); they cannot depend on merely mathematical properties such as those in (1).

But where in this division are we to place the congruence relations that exist between pairs of points? In the Axioms of Intuition of the *Critique,* Kant moves quickly from the rather sparse topological requirements he puts on the notion of an extensive magnitude (essentially the part-whole structure of the spatial continuum) to the claim that geometric truths are synthetic *a priori.* But he also emphasises that the possibility of establishing congruence relations is a basic premise of geometric science, the very one assumed as unproblematic in phoronomy. The topological structure of spatial intuition is not, on Kant's own admission, sufficient to schematise our spatial concepts: assertions of congruence can only be meaningful on the assumption of a constructive procedure involving comparison by means of displacement,[65] and thus Kant has no criterion at hand for distinguishing between the pure matter of phoronomy (which has the one characteristic of being movable in space) and the geometrical lines and points that are "drawn" and displaced when doing geometry. Because he allows that the constructive procedures underlying geometry involve operations with ideal point-systems, he lacks a clear-cut distinction between those spatial properties which belong to the groups (1) and (2) from above.

Considered on its own, Kant's theory of geometry and of the role of constructions in geometric demonstrations may or may not be defensible. But our concern, which is the same concern that comes to plague Helmholtz when he borrows these arguments, is not the internal consistency of Kant's theory of geometry. We are concerned with the consistency of this theory with the doctrine of empirical determinacy, and with the attempt to derive the form of force laws from this doctrine. My suggestion is that Kant cannot have both: if geometrical proofs involve constructive proce-

65 *Cf. Prolegomena* §12, 4.284.

dures, then there is no rigorous distinction between phoronomy and geometry. But then we must be prepared to countenance forces and force-laws that are far more diverse than those demanded by Kant. If on the other hand we insist on arguing from determinacy to the form of possible laws, then we will have to provide a theory of geometry that fuses it with phoronomy. This is, as I shall be arguing in the following chapters, the solution proposed by Helmholtz to resolve the dilemma.

3. Helmholtz on the Comprehension of Nature

At the conclusion of the last chapter, I argued that there is an implicit pressure on the epistemological status of geometry in Kant's theory of science. On the one hand, Kant himself insists that geometric propositions must be demonstrated by means of constructions. The propositions of geometry are revealed as synthetic *a priori* precisely because such procedures are required. If it were possible to derive the angle-sum of a triangle solely from the proposition that a triangle was a figure enclosed by three straight lines, then this proposition would be analytic. Whereas, according to Kant, no such proof is possible: in order to complete his proof, the geometer must extend the lines of the triangle, and must employ geometrical principles that refer to spatial properties that are external to the triangle, for instance the proposition that parallel lines never intersect. It is this very need to appeal to properties of the pure intuition of space (for instance its unboundedness) which demonstrates that space is an *a priori* form of appearances, and that the science which describes its properties is not a system of merely conceptual truths (*KrV* A716/B744).

But this insistence on the role of construction is at odds with another sequence of arguments in the *Metaphysical Foundations of Natural Science.* As we saw in our discussion in Chapter 2, Kant employs transcendental arguments in the *Metaphysical Foundations* in order to prove that forces are central, and that the parallelogram law of force composition is apodictic. These arguments accord a crucial role to what I called the principle of "positional determinacy." According to this principle, a spatial magnitude is only determinate once it is delimited by (at least) a pair of material points. Kant appeals to this principle to show not only that the action of a force must be understood relationally, so that we can only speak of a force as something which either attracts or repels the points along the line connecting them. He also claims that the magnitude of the force can vary only as a function of the same magnitude that the two points determine. In both cases, the argument rests on two suppositions: (1) the principle of positional determinacy, and (2) the requirement that all observable changes be determined by observable properties of the system, among which Kant includes the masses of its particles. Because (2) is demanded regulatively, and because (1) restricts the class of spatial magnitudes that qualify as observable in the sense of (2), it follows that forces must have the characteristics Kant demands. And once Kant has secured the centrality of force, he can couple it with a law of momentum conservation to prove that forces must sum according to the parallelogram law.

I suggested that Kant cannot consistently argue in both modes: if we are permitted in geometry to arbitrarily construct and displace mathematical figures in space, then any property thus definable can qualify as an admissible determinant of physical laws. Conversely, if we deny the latter possibility (which is the only approach consistent with the hypothesis that there is no absolute space), then we must be prepared to acknowledge that the constructive operations involved in geometrical demonstrations are not truly distinct from operations carried out with empirically given bodies. By what right can Kant maintain that the matter posited in the Phoronomy, which has no other properties beyond being movable in space, is distinct of the "matter" used in geometric constructions? It will not do to insist that such constructions do not involve material objects in the sense of the Phoronomy, for there must obviously be *something* that is moved from one region of space to another when a comparison of their lengths is required in a proof.[1]

Kant appears unaware of this difficulty in his system, however we can say with some certainty why he is led to introduce it. The argument of the *Metaphysical Foundations* is directed against Newton's understanding of his own laws of motion, according to which there is an absolute space, whose properties are codified in geometry. In consequence, Newton can maintain a distinction between absolute motion and rest, and he can also conceive of forces as independently existing quantities, whose direction and magnitude can be defined with respect to the directions and the metrical characteristics of absolute space. Kant, in contrast, rejects this interpretation, arguing that since absolute space is unobservable, it is not an object of possible experience, and thus has no place in the system of physical concepts. Thus he attempts to develop the core concepts of physics in the *Metaphysical Foundations* without any reference to an absolute space: to the extent that spatial magnitudes are involved in this development,

1 By the phrase "when a comparison of their lengths is required," I mean that when a statement, for instance the assumption of a *reductio*, posits the equal length of two line-segments, it also affirms implicitly that a comparison of these segments could be made. Such a comparison would involve the transport of a "mathematical," as opposed to material object. As I discuss in detail in Chapter 5, Helmholtz had already considered this notion at length in a manuscript on "general physical concepts" written some time before the publication of the *Conservation*.

they must all be delimited by empirically given points, even if these are themselves conceived as "pure empirical" constructions, as in phoronomy. But in making this move, Kant inadvertently puts the truths of geometry back in play. Which properties of the spatial manifold can be assumed to be determinate when a pair of points is given? By what right can we assume that it makes sense to speak of, for instance, two empirically given distances as having the same length? Does this not implicitly invoke the notion of a comparison of the two distances? And must not the comparison in question be one carried out with phoronomic instruments (with material rulers and compasses, as opposed to imagined ones), if the argument from determinacy is not to draw a blank?

In this chapter, I shall argue that Helmholtz was eventually forced to recognise just this dilemma in the course of his early physical research. Helmholtz took up Kant's arguments from the *Metaphysical Foundations* as the philosophical framework of his early memoir, the *Conservation of Energy.* He also argued for the *a priori* necessity of central forces on the basis of a transcendental principle, according to which nature must be "completely comprehensible" (*vollständig begreiflich*). And he adopted this transcendental method in order to neutralise theories of electromagnetism that were inadmissible, on his view, because they involved indeterminate magnitudes. In the course of defending the *Conservation* against sophisticated criticisms levelled by Clausius, Lipschitz and others, Helmholtz dug himself deeper into Kant's trench. According to these authors, Helmholtz had been overly hasty in deriving the centrality of force from his transcendental postulates. In particular, it was perfectly well conceivable that there might be forces in nature that satisfied Helmholtz's energy conservation principle without being central. In a reply to Clausius from 1854, Helmholtz sought to repel the latter's advance by insisting that all spatial magnitudes introduced into a theory had to be "relations among real things." He argued that this constraint disqualified forces of the sort imagined by Clausius, since the latter were purely mathematical, in that they were definable only by means of coordinate systems drawn arbitrarily "on paper." These arguments, I shall suggest in the last section of this chapter, put Helmholtz in the position of having to explain how and why we can speak of relations of congruence among material systems without appealing to purely mathematical coordinate systems. For Helmholtz's formulation of the energy principle assumed that one could meaningfully speak of the same system being in a congruent state at a later moment in time. If in fact every coordinate system has to refer to relations among real objects, as Helmholtz insisted, this statement alone implied the existence of some non-mathematical coordinate system with reference to which one could meaningfully say that energy was conserved.

What I will not suggest in the following is that Helmholtz, in 1854, took his reasoning to imply that geometry was an empirical science. On the contrary, I take it that his work in an entirely separate field, namely the sense-physiology of colour-perception, led him to pose the question in this form. But this work, as we shall see in the next chapter, was being carried out simultaneously to the debate with Clausius. My thesis, for while I shall argue in greater detail in Chapters 4 and 5, is therefore that Helmholtz was gaining insight into the possibility of empiricising spatial relations in his sense-physiological research at exactly the moment that he was insisting on the empirical character of all physical magnitudes in the context of the debate of energy conservation and electrodynamics. The attempt to provide a transcendental proof of the empirical validity of Euclidean geometry—a project that is scarcely comprehensible outside the context I am describing—was the upshot of this dual research programme. Helmholtz sought to offer such a proof at just that point in his career when he left physiology to return to physics.

Thus my task in this chapter will not be to show that Helmholtz doubted the validity of Euclidean geometry during the phase I am describing. The result of our discussion should be only that outlined at the end of the previous chapter: I shall demonstrate that the implicit tensions in Kant's system became evident to Helmholtz precisely because he carried through on Kant's arguments with philosophical and mathematical rigour. This analysis, and the ensuing debate, took him to the brink of a divide which he eventually crossed in 1868, when the first papers on geometry were published.

(a) Helmholtz's Arguments for Force Centrality in the *Conservation of Energy*

In his 1847 *On the Conservation of Energy*, Helmholtz had argued that the principle of *vis viva* conservation was equivalent to the hypothesis that all forces of nature were central forces holding between mass-points. But, he claimed, this latter hypothesis was necessary. For science aims at providing a complete description of the natural world, and this requires that all phenomena it treats of must be empirically determinate. Helmholtz thought that he could derive both of the two defining characteristics of central forces from restrictions on the form of those laws and phenomena that were compatible with this requirement: first, the forces had to be directed along the line connecting the mass-points, for this was the only spatial magnitude determined by the two points; second, in order that science be maximally unified, their intensity also had to be a function of this one magnitude. Both of these demands supposedly followed from the essential *in*determinacy of spatial relations, that is to say from their relativity, which will be discussed in detail in the following. Thus the characteristics of central forces derive from conditions on the *determinate representability* of motions and forces.

When the *Conservation* was republished in the first volume of his collected works in 1882, Helmholtz appended a series of supplementary, and on some points critical remarks on his earlier positions. Here, he modified both of the above claims. He conceded that his aprioristic argument for the necessity of central forces was incorrect, and thus that it represented at most an empirical generalisation. Furthermore, he admitted that this empirical generalisation had itself been called into question by current theories of electromagnetism. This state of affairs pointed, in his view, to a critical tension in the state of physical theory. The epistemological privilege he had ascribed to central forces remained unchanged, for the philosophical argument from determinacy was not wholly mistaken. But the new theories of Weber and others had met with undeniable success. These theories thereby threatened to force physics to abandon the goal of providing a completely determinate theory of nature, and therefore also the hope of completing natural science. Thus there was a conflict between empirical adequacy and epistemological coherence. In order to clarify this dilemma, I shall briefly contrast his early account the matter with his reappraisal from 1882.

(i) The Philosophical Argument

In his original presentation in the *Conservation* of 1847, Helmholtz had maintained that forces had to be central because this postulate followed from the regulative demand that nature be completely comprehensible (*vollständig begreiflich*). This demand would be fully satisfied just in case we had articulated laws allowing us to predict the (future) behaviour of each natural system whose material properties were fully known to us in the present. In such a complete science, our knowledge of the physical world would cleave neatly into two parts: (1) temporally invariant general laws and (2) descriptions of particular physical systems. In a pure mathematical theory of the sort Helmholtz required, the properties allowed to these systems are restricted to the masses and the positions of their material points, along with their velocities and the forces holding between them. The exact nature of these forces will depend on the kinds of matter we are dealing with. Indeed, the differences among kinds of matter will, in such a complete theory, be represented solely as differences in their masses and in the forces the various matters generate.[2]

According to Helmholtz, forces are the causes of changes in the motions of the mass-points. But the intensity of such forces can evidently vary with time. Thus a final reduction of physical phenomena to time-independent laws will require our identifying the fundamental invariant forces characterising the various species of matter. These fundamental forces will have to be described by functions that depend on empirically determinate properties only, and which, in particular, do not depend on time. But since the mass of material particles is assumed to be constant, the empirically determinate properties spoken of here can consist only in the positional relations among the points.

This result implied, according to Helmholtz, the dual connection between force and position that I outlined above: on the one hand, forces are the ultimate causes of observed changes in position; on the other hand, if these forces themselves are subject to change, the latter changes must depend on these same positions. Helmholtz then applies what I have previously called the principle of positional determinacy, in order to prove centrality. He demands that any magnitudes used to characterise the motion of a system be defined only in terms of the relative positions of its mass-points. Such magnitudes, which include directions and distances, we will call the *internal properties* of the system. While directions and distances in absolute space may enter into our mathematical description of the system, the external properties cannot be objects of experience, and thus they are physically indeterminate.

As we have seen, the principle of positional determinacy restricts the range of motive concepts that can be applied to systems. A single point in space determines no spatial magnitudes at all, thus it cannot be said to undergo motion, let alone accelerated or forced motion. When two points are given there can indeed be relative motion, but only one direction and only one magnitude are determinate, and so on. Helmholtz, al-

2 *Cf.* Helmholtz's manuscript from the period before the *Erhaltung* reproduced in (Königsberger 1903), pp. 126–138, p. 131.

most without comment, restricts the results concerning the positional dependence of forces to the elements of a two-point system, and thereby derives the two elements of centrality from positional determinacy: (1) Force intensities must be functions of the distance determined by the two mass-points. For if this were not the case, then changes in the forces acting among the points (changes in the causes of change), would depend on non-observable, and thus experientially indeterminate properties of the system. And this would mean in turn that nature was not completely comprehensible in the desired sense. Furthermore, (2) the only observable effect of the force acting between two mass-points can be to alter their distance. Finally, by (1) and (2) it follows that all forces in nature must be central forces, and the demand that nature be completely comprehensible entails that all forces are central.

This transcendental deduction of central forces in the *Conservation* is followed by a mathematical demonstration of the equivalence of the postulate of force centrality to the principle of the conservation of *vis viva.* The latter principle, in Helmholtz's version, states that whenever a system is in the same state—that is to say whenever all the internal properties of the system are the same—the kinetic energy of the system is the same as well, whatever the path followed by the system in the intervening time. Helmholtz's proof, which we shall examine in greater detail below, contained a number of flaws, one of which was shared with the philosophical deduction just outlined. For this deduction assumes that we can draw conclusions about the nature of fundamental forces by restricting ourselves to two-point systems and then applying the principle of positional determinacy. Helmholtz did the same in his mathematical demonstration of the equivalence of centrality to *vis viva* conservation. In addition, he overlooked the possibility that forces could depend on the velocities and accelerations of masses. But let us set these difficulties aside for the moment, so that we can get a clear view of the aim of these two proofs.

Suppose that both the philosophical deduction of centrality from the comprehensibility of nature *and* the mathematical demonstration of the equivalence of centrality to *vis viva* conservation had been valid. What exactly would Helmholtz have shown? We have,

I. comprehensibility of nature $\Rightarrow$ force centrality

II. force centrality $\Leftrightarrow$ conservation of *vis viva*

Thus it follows that,

III. comprehensibility of nature $\Rightarrow$ conservation of *vis viva*

Proposition III expresses a Leibnizian intuition, namely that if it were possible to construct a perpetuum mobile (if the same system, in the same internal state, had a different kinetic energy) this would violate the principle of sufficient reason. A theory implying the possibility of perpetual motion is not merely empirically false, it is inadmissible on epistemological, if not metaphysical grounds. But the implication that expresses this supposed truth, however persuasive it may be intuitively, is hardly rigorous enough to do physics. Whereas Helmholtz's deduction of force centrality from the comprehensibility of nature is not only philosophically persuasive, but it also yields a proposition with a precise physical content. Furthermore, Helmholtz has also refined the antecedent to the implication. His definition of the comprehensibility of nature makes specific demands on the forms of laws and of phenomena, and he believes he has shown that these specific demands entail a result which, in turn, entails energy conservation. Helmholtz's analysis is therefore not merely of philosophical interest—it can be, and it was used by Helmholtz to argue philosophically against physical theories that were not in agreement with the requirement of force centrality. But for this very reason, the coherence of the proofs and the modal status of the premisses are of decisive importance. As we have seen, the proofs of both I and II contain additional premisses, and these must have at least as much of a claim to *a priori* validity as the principle of the comprehensibility of nature if the philosophical argument is to do any physical work.

In moving from the claim that the forces within a system must depend on the positions of the points alone to the conclusion that they must be central, Helmholtz appealed implicitly to two such premisses: (1) what Olivier Darrigol has called the principle of decomposition,[3] and (2) the principle of positional determinacy. According to (1), the force acting on a single point in a system is the (geometric) sum of the forces deriving from the other points in the system. In order to characterise any one of these forces, we may ignore the positions and motions of the other masses in the system, and confine our attention to just these two points. Only by invoking this principle can Helmholtz prove that the intensity of a force holding between two arbitrary points varies with *their* positions only. According to (2), neither directions nor distances are determinate unless they are delimited by empirically given points.

3 (Darrigol 1994), p. 217. Helmholtz's principle is "a particular case of what I shall call the *principle of decomposition*, according to which all actions in nature must be resolved into actions involving only two elements of volume."

Thus only spatial properties defined with reference to the points involved in the system under consideration can be employed to determine its future motion. By invoking (2), Helmholtz is able to argue that "dependence on position" in the case of two points can mean only "dependence on distance."

As should be evident from our discussion in the last chapter, Helmholtz's arguments throughout these introductory sections are closely related to those used by Kant in his *Metaphysical Foundations of Natural Science.* The introductory section of the *Conservation* argues for the logical connection between a systemic principle—the complete comprehensibility of nature—and what would appear to be a contingent law of natural science, namely the conservation of energy. But one should not be misled. It is not a proof that force centrality must be, or probably is to be found in nature. It does not demonstrate the necessity of central forces, if one means by this either their logical or mathematical necessity, or indeed a metaphysical necessity deriving from first principles. There is no guarantee that nature *is* completely comprehensible, nor indeed should one conclude from such a proof that one has good reasons for *believing* that forces are, in fact, central. The point is purely methodological: *if* we approach nature with the intention of producing a complete set of invariant laws, *then* it follows that only certain kinds of laws are going to work, in the sense of being adequate to the task we have set ourselves. The necessity in question is a *regulative* necessity.

Helmholtz purports to have shown that these laws will involve central forces, because any other sort of description will fail to realise our aims. But even if this is a methodological, and not a metaphysical thesis, it still involves a strong negative claim, namely that certain kinds of laws are non-starters. Furthermore, the proof involves more premisses than just the regulative principle. Finally, the negative claim is just as little an empirical claim as the positive conclusion of centrality. And so the premisses involved cannot be merely empirical, for otherwise we should have drawn a conclusion concerning the necessary form of scientific theories from propositions that are possibly false. This would mean that the proof in question was no proof at all. Helmholtz therefore attempts to argue for these principles by suggesting that they derive from conditions on the possibility of representing nature *at all*, and in this sense they can and should be called transcendental in Kant's sense of the term. For instance, according to Helmholtz, forces can depend only on the relative distances holding between points because (1) regulatively, we must seek to simplify forces by regarding them as characterised by time-independent functions, and because (2) constitutively, the distances that are candidates for the parameters of these functions must be possible intuitions, i. e. spatially determinate magnitudes.

Helmholtz does not explicitly state the principle of decomposition in the *Conservation* (he does so only in a later article from 1854),[4] but he does in effect apply it there when making the transition from the actions of forces in complex systems to the actions of the forces holding between pairs of their components. He sees this step as unproblematic because it apparently follows from his definition of a force: a force is the cause of a change in position, and such a change cannot be observed unless there is a second point relative to which the change takes place. Thus motive force is "to be defined as the striving of two masses to change their relative positions."[5] But the reasoning is fallacious: because a change in position can be observed only when there are "at least two points," Helmholtz concludes that all complex systems must be resolved into sums of two-point systems, i. e., into cases in which there are *at most* two points. As we shall see in a moment, Helmholtz was forced to retrench from this part of his analysis in 1882. But he did not step down from the principle of positional determinacy, so much as use it as the starting point for his later work on geometry.

By claiming that the supplementary assumptions in his proofs derive from conditions on the intuitive representation of motions and forces, Helmholtz secures a connection between the comprehensibility of nature and the central force hypothesis that he derives from it. The latter hypothesis thereby inherits the status of a regulative principle: it could be frustrated by experience, but in such an event, no alternative, comprehensive theory of nature would be possible. Thus we must seek to formulate theories by means of central force laws. Given the structure of this argument, one might conclude that the errors in his deduction would have led Helmholtz to abandon his commitment to the central force hypothesis. But even in 1882, after he had acknowledged the formal errors and well after he had recognised that both of the transcendental principles might be empirically false, he continues to argue that the central force hypothesis is epistemologically privileged. This fact is at first glance puzzling; however, Helmholtz's position is just as coherent, or just as contradictory, as his later views on geometry. For there as well, he argued for an interpretation of physical theory in which certain principles (the axioms

4 (Helmholtz 1854)
5 (Helmholtz 1996), p. 6.

of geometry) were *necessary* for the description of physical phenomena, even though the evidence for these principles was of inductive origin. The distinction between regulative and constitutive premisses suggests the direction in which Helmholtz retrenched: regulative principles point at methods of representation that are systemically preferable, we might say; whereas in saying that a principle is constitutive, I make a stronger claim. Constitutive principles are not *preferable* to their alternatives; rather, the alternatives are incoherent, they are not alternatives at all. But if one demotes a principle from constitutive to regulative status, one can continue to employ it as a premise in a transcendental argument. Even at that date in his career when Helmholtz presented himself as an ardent empiricist, he continues to ascribe a privileged role to central forces—they continue to be necessary for the construction of comprehensive physical theories. But this requirement is no longer a necessary truth in the strict, constitutive sense of that term.

When the *Conservation* was reprinted in his 1882 collection of physical papers,[6] Helmholtz reinterpreted his early arguments along just these lines. He gave a detailed comment on the passage where he had argued that forces had to be resolved into the forces acting between point-masses if science was to be comprehensive and determinate. He now distinguished between two senses in which this claim might be true. It might mean that (1) the motion of every point in the system had to be determinable once the forces present in the system were known (forces act on point-masses). And it could also mean that (2) the force acting on each point could be decomposed into the forces emanating from all the other points in the system (forces emanate from point masses). The first interpretation was indeed epistemologically necessary. For if there were individual points in a system whose motion was not determined by the forces acting within the system, then certain motions of the system would indeed fail to fall under general physical laws. The second interpretation is our principle of decomposition, whose importance Helmholtz first isolated in his 1854 "Reply to Clausius," but which he then viewed as unproblematic. This principle, Helmholtz now admitted, was purely contingent, and it represented "the real content of Newton's second axiom."[7] According to Helmholtz, it states that the acceleration undergone by a particle

6 This was the first volume of his *Wissenschaftliche Abhandlungen*, (Helmholtz 1882). The comments are dated 1881, whereas the volume was actually printed in 1882.

7 (Helmholtz 1996), p. 54.

when several causes act at once is equal to the geometric sum of the accelerations it undergoes when these same causes act severally. That is to say, Newton's parallelogram of forces can be applied in reverse in order to resolve the acceleration onto components directed towards the other points in the system.

This way of stating the principle of decomposition makes clear its importance to the proof of central forces. For if one assumes—as both Helmholtz and Kant before him had done—that force is *by definition* the capacity of one body to alter its position relative to another, then this principle can easily seem apodictically true. The error in this reasoning lies in assuming that the change in the position of the two points referred to in the definition of force *must* be a change in the distance along the straight line connecting them, as is the case in such a simple system. And so it may seem as though the principle of decomposition follows logically from the principle of positional determinacy and the definition of force. But so long as the system is complex enough to define other directional and metrical relations, this conclusion no longer follows. Thus, as Helmholtz is now forced to admit, the principle states an empirical truth. Nevertheless, he contends that all theories that do not assume it have also been found to contradict the principle of energy conservation, and indeed the law of equal action and reaction. And the same holds true of attempts to do without the assumption that forces are determinate "once the positions of the masses are completely given."[8] Finally, to suggest that forces might be "made dependent on an absolute motion, that is to say on an alteration of the relation of a mass to something that could never the object of a possible perception" is to abandon all hope of "a complete solution of natural scientific problems."[9]

In other words, Helmholtz does not conclude that his original analysis was wholly mistaken. On the contrary, he suggests that the conflict between his earlier philosophical views and the state of electrodynamic theory points to a fundamental tension. Implicitly referring to Weber's electrodynamic theory, he maintains that (1) theories that allow forces that are not central threaten the unambiguousness (*Eindeutigkeit*) of electrical theory. And to assume that, as in a field-theory, (2) the intensity of a force might depend on its velocity relative to absolute space, is to abdicate the fundamental responsibility of physics, which is to comprehend nature completely. Although it is not logically incoherent to deny the necessity

8 (Helmholtz 1996), p. 54.
9 (Helmholtz 1996), p. 55.

of central forces, to do so is to deny the validity of hypotheses which, while contingent, nevertheless remain conditions on the determinacy and completeness of natural science. And this step can only be taken, Helmholtz maintained, once it is clear that all other alternatives have been exhausted.

Both of Helmholtz's retractions in 1882 derived from criticisms of his book that had been raised early on by Clausius and Lipschitz.[10] Lipschitz's objection centred on Helmholtz's mathematical proof of the equivalence of force centrality with energy conservation. He showed that velocity- and acceleration-dependent forces were consistent with Helmholtz's definition of conservation, and thus that exclusively positional dependence did not follow without further assumptions. Clausius raised a number of objections to the text, of which only one is of immediate concern here, namely his contention that the forces acting on material points did not have to be central merely because they were position-dependent. Helmholtz had made this inference from positional dependence to centrality twice: first in the philosophical deduction outlined above, and again in his mathematical demonstration of the equivalence of centrality and conservation. We have already examined Helmholtz's reasoning in the philosophical proof; however, in order to understand Clausius's objection and Helmholtz's response, it is worth examining the mathematical demonstration in greater detail. For it is in attempting to salvage this argument from Clausius's criticisms that Helmholtz first contends that the possibility of establishing congruence relations is a condition for the physical application of a theory, pointing out that this condition is generally overlooked.

(ii) The Mathematical Argument

This mathematical demonstration contained two parts, each of which proved one of the two implications involved in the equivalence of force centrality to *vis viva* conservation. Obviously it was not difficult to demonstrate that centrality implies conservation (that central forces are conservative), so we shall focus on the problematic implication, namely that centrality follows from *vis viva* conservation. Helmholtz's proof runs as follows. He first defines *vis viva* conservation as the proposition that "when an arbitrary number of mass-points moves under the influ-

10 See (Bevilacqua 1993), p. 313, (Darrigol 1994) p. 221, (Bevilacqua 1994).

ence of only those forces that they exert on each other … the sum of the living force of all of these is the same at all points in time in which they adopt the same relative position to one another."[11] This definition implies the existence of a potential

$$(1)\quad d(q^2) = \frac{d(q^2)}{dx}dx + \frac{d(q^2)}{dy}dy + \frac{d(q^2)}{dz}dz$$

where q is the tangential velocity of the mass-point relative to a system A, and where the coordinates are defined relative to that system. If u, v, and w are then the components of the motion along the axes x, y, and z, we obtain the following expressions for the component forces on the point

$$X = m\frac{du}{dt},\quad Y = m\frac{dv}{dt},\quad Z = m\frac{dw}{dt}$$

from which it follows immediately that

$$du = \frac{Xdt}{m}, \text{ etc.}$$

Now, since $q^2 = u^2 + v^2 + w^2$, thus $d(q^2) = 2udu + 2vdv + 2wdw$, and since

$$u = \frac{dx}{dt}, \text{ etc.}$$

we can substitute for each term $2udu$ one of the form

$$\frac{2X}{m}dx, \text{ etc.}$$

in order to derive

$$(2)\quad d(q^2) = \frac{2X}{m}dx + \frac{2Y}{m}dy + \frac{2Z}{m}dz$$

Helmholtz thought that he could immediate derive from (1) and (2) that,

$$(3)\quad \frac{d(q^2)}{dx} = \frac{2X}{m},\quad \frac{d(q^2)}{dy} = \frac{2Y}{m},\quad \text{and}\quad \frac{d(q^2)}{dz} = \frac{2Z}{m}$$

And then, on the assumption that q^2 is a function only of the coordinates x, y, and z, it follows that the component forces X, Y, and Z are also functions only of these coordinates, and thus by definition they are functions of the relative positions of the mass-point to the system A. Nevertheless, in deriving (3), Helmholtz overlooked the possibility identified by Lipschitz, namely that the force along each axis could depend on the velocity and acceleration of the particle. But if we grant him that simplification,

11 (Helmholtz 1996), p. 9.

Helmholtz has shown that the force acting on each individual point is a function of the position of the point relative to the system as a whole.[12]

There remains one last step to the proof, for the positional dependence of the force relative to the whole system is not the same as centrality. To complete his proof, Helmholtz needed to get from this sense of positional dependence to the narrower, two-point case. Tacitly invoking the principle of decomposition, he restricted this result to the case of two material points. Since by hypothesis, the coordinate system being used is determined by the mass-points involved in the system, in the two-point case this coordinate system can be reduced to the single dimension determined by the two points. Thus, (4) the energetic state of the one particle is a function only of its distance from the second. It then follows trivially that (5) the positions in which the first point is in the same energetic state form concentric shells about the second point, and thus that the force acting on the first is directed toward the second. By (4) and (5) the force is a central force.

In his critical report on the *Conservation*, Clausius objected that this part of the proof simply begged the question. For Helmholtz had merely assumed that which needed to be demonstrated, namely that the intensity of the force holding between two points was a function of their distance.[13] The derivation of its direction from this postulate was then a trivial business. Furthermore, Clausius showed that one could describe force functions that both satisfied *vis viva* conservation and were position-dependent, without being central. In a long reply to this and other objections of Clausius, Helmholtz defended his procedure on the following grounds.[14]

There were two fundamental assumptions involved in his earlier proof, he argued. The first of these was that the forces acting on a given point in a system could be decomposed by means of geometrical addition, that is to say by employing Newton's parallelogram of forces. But this principle, he contended, was independent of Clausius's objection. Their main difference of opinion, he went on, concerned the latter's supposition that we can coherently imagine forces holding between

12 In the 1882 comments, Hemholtz suggests adding the requirements of equal action and reaction, and of the reducibility of forces onto point-masses (i. e. the principle of decomposition) in order to rule out forces of the sort admitted by Lipschitz. Of course, that is just to insist on principles that entail centrality. They are, incidentally, the same principles that Kant requires in the *MFNS.*

13 (Clausius 1853), pp. 574–578.

14 (Helmholtz 1854), pp. 81–83.

masses that are not functions of their relative position, that is to say of their distance. In the second part of his proof, Helmholtz argued, he had merely confined his more general result, namely that force was a function position alone, a result which Clausius accepted, to its minimal case. Under the assumption that it was only meaningful to speak of *relative* positions, in other words to make reference to coordinate systems that were empirically given, it followed that in the case of a two-point system, dependence on position was synonymous with dependence on distance. From this one could derive the requisite direction of the force, and therefore its centrality. Clausius's non-central forces, by contrast, could be defined only in terms of an absolute coordinate system. But this meant supposing that they were determined by something which could never be an object of possible experience, namely absolute space.

In order to understand this emphasis on possible experience, one must recall Helmholtz's insistence on the completeness and determinacy of natural science. As I indicated previously, it was an essential tenet of the *Conservation of Energy* that the ultimate causes in nature are unchanging, even though the forces we find in nature do quite obviously change in intensity. But this demand can be satisfied if these changes in intensity are describable as law-like functions of spatial magnitudes. For in this case, the intensity of the force depends wholly on properties of a system that are independent of the time. For while it is true that these properties change *in* time, we do not need to take account of time (i. e. by introducing it as an independent variable) in order to describe the changes in force intensity.

But this solution will ensure complete comprehensibility only if these positional properties are themselves empirically determinate. And since space is undifferentiated, we can speak of determinate spatial magnitudes only when there are empirical points demarcating intervals in space. Thus, in the case of an isolated system, the forces will have to vary in accordance with the magnitudes determined by the relative positions of the points in the system. In other words, changes in the relative positions of mass-points in space are not only the experiential *consequences* of the actions of forces. They are also the only possible *determinants* of changes in the forces' intensities.[15] The problem with the deviant forces that Clausius imagines is that they would have to depend on features of reality that are not possible experiences. They are, in consequence, experientially indeter-

15 *Cf.* (Heidelberger 1993), p. 470.

minate: if the intensities of forces so defined change with time, these changes depend on factors that cannot be empirically identified. Claiming that they are functionally dependent on changes relative to an absolute coordinate system is thus an empty gesture. For in admitting non-experiential determinations of physical values, we would in effect be admitting transcendent elements in our theory.

Nevertheless, as I have already suggested, Helmholtz's reply to Clausius cleaves the problem at the wrong joint. Suppose we allow him proposition (3) from his proof, which states that the energetic state of a particle depends only on its position. And suppose we agree that references to absolute coordinate systems are to be disallowed because they are transcendent. Does the proof then go through? No, because the application of the principle of decomposition effectively eliminates determinate spatial relations which *could* be used to define non-central forces. That Helmholtz apparently failed to see this, while devoting several pages of his "Reply" to a deduction of the mathematical consequences of his fallacious philosophical argument provides ample evidence of what he admitted in 1882, namely that the philosophical portions of the *Conservation* are "influenced by epistemological views of Kant that are stronger than what I would be prepared to accept today."[16]

Helmholtz expands on this statement by explaining how his views on causality and matter have changed since his early years; however, the connections to Kant's philosophy of physics run deeper than this, and they include the ongoing confusion concerning the epistemological status of the principle of decomposition. Helmholtz, like Kant, had acted as if the latter principle were in some sense entailed by the principle of positional determinacy, for he repeatedly suggests that centrality is entailed by the requirement that all basic magnitudes be empirically determinate. Now, as we have just seen, *if* the principle of decomposition holds, positional determinacy entails centrality. And decomposition itself might appear to follow from Kant's definition of force as the cause of a change in the relation between two mass-points. For one might reason as follows: *if* a force is a relational property of a pair of points, then it should be fully determined by the properties of that pair of points (by their masses and positions). Thus whenever a point is subject to the actions of forces emanating from several masses, each force must be conceived as unaffected by the presence of the other masses. This means supposing that the positions of other points in a system do not add to the properties of the point-pair,

16 (Helmholtz 1996), p. 53.

because they are "external" to the pair. And that might in turn appear to be nothing more than a special case of the relativity of spatial relations. Since it is illegitimate to refer to absolute space in characterising the position of two points, it might also seem to be illegitimate to refer to any other points but the two under consideration, for these are irrelevant to characterising their position *relative to one another.* But this reasoning is obviously circular. For the position of a pair of points relative to some third point can very well be seen as a property of that pair, and thus as something that might determine a force that nevertheless *acts* only between the pair. In sum, if decomposition holds, then forces determine, and are determined by, the distances of pairs of points; whereas assuming that only these distances are relevant to characterising the actions and variation of forces is in effect to assume decompositionality.

(b) Helmholtz's Later Criticisms of his Determinacy Argument

As should be clear from the previous discussion, Helmholtz's reply to Clausius, and indeed his insistence on the epistemological priority of positional relations, is nothing other than a transcendental argument. Clausius's arbitrary forces are ruled out because they can be given only a hypothetical, "mathematical" definition, whereas Helmholtz will allow only forces agreeing to an epistemological demand: Not only must their effects be determinate, but they themselves should be *determined,* meaning that their intensities must depend only on spatial relations which are themselves, in turn, determinate. Helmholtz's arguments therefore contain two distinct appeals to the notion of logical determination that are easily confused. The first is the more straightforward, "downward" demand that the extension of a concept be determinate. It should be defined precisely enough that we may say of each object whether or not the concept applies to it. For instance, one might argue that a precise definition of the concept of force will also have to contain some reference to an inertial frame, because without this specification, the concept of an accelerated motion, and thus the concept of the cause of an acceleration is indeterminate. But Helmholtz also appeals to logical determination in a second, "upward" sense: each concept must be *determined,* in that it should be seen as a specific case of a higher-order concept. The specific forces that cause changes in motion should be seen as positional determinations of a single force; furthermore, this single force should emanate from a material point, and it should itself be a member of the family of "basic forces" that define the various kinds of matter.

These upward and downward demands correspond quite well to the regulative and constitutive constraints placed by Kant on the schemata of pure empirical concepts in the *Metaphysical Foundations.* As we saw in the last chapter, Kant requires each high-level concept in a conceptual hierarchy to subsume, or to exclude, any lower-level concepts if it is to qualify as determinate. And the same holds conversely: for every lower-level concept, there must be a determinate answer to the question of whether it falls under a higher-level one. In a pure empirical system of concepts, matters are complicated by the need to fix these logical relations *a priori.* For it is only at the base level—that of phoronomic concepts—that we can provide a class of subsumable entities by means of mathematical constructions. These constructions are, moreover, still equivocal: one and the same construction of the motion of a pair of points could fall under several, apparently distinct phoronomical concepts, depending on which of the pair is considered to be at rest. It is just this indeterminacy that Kant believes will be eliminated by the introduction of dynamical and mechanical principles. Nevertheless, the downward, constitutive demand still sets limits on the possible schemata of physical concepts, and these limits derive from the possibilities of spatial constructions. They are the same constitutive conditions that yield the synthetic *a priori* truths of geometry.

The upward, or regulative requirement constrains the class of admissible forces, that is to say, the concepts *under which* the phoronomic concepts and their corresponding intuitions will be subsumed. Forces are to be represented as magnitudes, as they must be if they are to be integrated in the apodictic core of a "proper science." As such, they must conform to mathematical laws—they must, in our terminology, satisfy additivity axioms. Thus a central aim of the *Metaphysical Foundations* is to provide *a priori* representations of forces from which their laws of composition follow necessarily. Newton must introduce the notion of independently existing forces in absolute space in order to prove the law of force addition. Kant cannot admit such a demonstration, however, because he holds the notion of absolute space to be experientially transcendent. The magnitudinal concept of force must therefore be schematised in such a way as to eliminate all reference to the external properties of the system. Kant mistakenly thought that he could satisfy this demand by showing that the actions of forces could only be described in terms of attraction and repulsion along the lines connecting pairs of points. Newton's law of force composition would thereby be proved apodictically, and no indeterminate concepts are introduced into the metaphysics of nature, as is required in a complete rational science.

As should be evident from the preceding discussion, Helmholtz's reasoning is close to Kant's not only in its general aims, but also in the details. Common to both men is the belief that *a priori* constraints on the intuitive "construction" of the concept of force disqualify absolute forces—forces whose magnitudes and directions are, as it were, anchored in absolute space. Helmholtz thought that he could employ such arguments in order to invalidate competing electrodynamic theories: if a theory made appeal to forces that were not experientially determinate, then it could be rejected. To be experientially determinate, a force would have to be central, for all other options amounted to reifying or absolutising forces. In Kantian terminology, both Kant and Helmholtz object to definitions (empirical schemata) of force that involve reference either to motion, or to absolute space, because they invoke features of reality that are *in principle* indeterminate. In the first case, we have a dependency on time, and in the second, a dependence on absolute space.

As I have already explained, in the course of eliminating these transcendent dependencies, Helmholtz employed two principles to derive of centrality from the requirement that nature be completely comprehensible: the principle of decomposition, and the principle of positional determinacy. Helmholtz needed decomposition in order to get centrality out of positional determinacy, for without it, he could not reduce complex systems to their component pairs. And this difficulty was precisely that which Clausius homed in on, even though Helmholtz did not at first grasp the full import of his opponent's attack. Clausius's critique, we may recall, was directed at Helmholtz's mathematical derivation of the equivalence of force centrality to his principle of *vis viva* conservation. The tricky bit was to prove one of the two implications making up the equivalence, namely:

conservation of *vis viva* $\Rightarrow$ postulate of central forces

To prove this proposition in the *Conservation*, Helmholtz applied the principle of decomposition (without explicitly flagging this step) to reduce the case of a complex system to a conjunction of two-point systems. Invoking the principle of positional determinacy, he claimed that the magnitude of the force could depend only on the distance between the two points. He then proved that the force also had to be directed along the line connecting the points. Clausius objected in turn that this was a *petitio:* Helmholtz assumed one half of centrality (dependence on distance) in order to prove the other (directionality).[17] Now, as we just saw, Helmholtz assumes the principle of decomposition in the original text of the *Conservation* without special mention or justification. However, in his "Reply" to Clausius, he does acknowledge his use of it, and reformulates his arguments in order to distinguish clearly between the two assumptions he had tacitly made before: (1) the kinetic energy of a system is the same *whenever the system is in the same (relative) state* (a question of positional determinacy), and (2) the force acting between any two points of the system is independent of the other points in the system (the principle of decomposition).

Helmholtz lays special emphasis on the notion of "same relative position," observing that this concept "has not been applied by all mechanists who have made use of this principle [the conservation of energy], but it is obviously necessary to its physical application." He suggests that we can define it as follows: "Movable points have the same relative position to one another whenever a coordinate system can be constructed in which all their coordinates have the same corresponding values."[18] From this definition he concludes immediately that the two points have the same relative position to one another whenever they are at the same distance. This immediately raises the question of *how we know* they are at the same distance. For, according to positional determinacy, the only spatial magnitudes defined are those determined by the material points in the system in question. As the phrase "constructing a coordinate system" implies, there must be some independent criterion for determining that the two points are at the same distance, and this will have to involve other material points.

Helmholtz concluded his 1882 comments on the opening sections of the *Conservation* by observing that *both* the principle of decomposition, *and* the result that forces depend on position only had been called into question in electrodynamics. As he had himself maintained in 1872, "Weber's hypothesis concerning electrical forces is the first, at least partially successful attempt to base an explanation of a class of phenomena … on the assumption of forces that depend not only on the position of mass-points, but also on their motion."[19] Furthermore, as I mentioned briefly above, the consistency of such forces with his conservation princi-

17 (Clausius 1853), pp. 574–578.
18 (Helmholtz 1854), p. 83.
19 (Helmholtz 1872), p. 645.

ple was an objection raised early on by Lipschitz, whom Helmholtz had been unable to refute as he had Clausius. Finally, as he himself admitted, he had been wrong to assume that these principles could be proven by means of *a priori* arguments. And yet, despite his recognition of the contingency of the principles he had used in his early work, Helmholtz continued to maintain that they were conditions for the "univocity and determinacy" (*Eindeutigkeit und Bestimmtheit*) of physical theory. In other words, Helmholtz retained the transcendental argumentation he had borrowed from Kant—certain physical principles are singled out by virtue of their making possible a determinate description of reality—all while relativising the status of these principles.

(c) Empirical Determinacy and Geometry

Even though Helmholtz's notebooks in the period before the *Conservation*[20] reveal that he had been concerned with the empirical conditions on spatial measurement at an earlier phase, he first puts these reflections to epistemological work in his reply to Clausius's objections to the *Conservation.* He does this by explicating the meaning of the term "relative position" that he had used in the *Conservation* to single out those states of a system that are equivalent from the point of view of the conservation law. The original formulation of the principle in the *Conservation* was the following:

> When an arbitrary number of mass-points move under the influence of only those forces that they exert on each other, *or that are directed at fixed centres,* then the sum of the living force of all of these is the same at all points in time in which they adopt the same relative position to one another *and to the possibly given fixed centres,* whatever their paths and speeds in the intervening time may have been.[21]

20 Quoted in (Königsberger 1903), pp. 126–138. Königsberger does not give an exact source or date for the manuscript he reproduces.

21 "Wenn sich eine beliebige Zahl beweglicher Massenpunkte nur unter dem Einfluss solcher Kräfte bewegt, welche sie selbst gegen einander ausüben, *oder welche gegen feste Centren gerichtet sind:* so ist die Summe der lebendigen Kräfte aller zusammen genommen zu allen Zeitpunkten dieselbe, in welchen alle Punkte dieselben relativen Lagen gegen einander *und gegen die etwa vorhandenen festen Centren* einnehmen, wie auch ihre Bahnen und Geschwindigkeiten in der Zwischenzeit gewesen sein mögen." (Helmholtz 1996), p. 9, my emphasis. The exact wording of the new definition in (Helmholtz 1854), pp. 82–83, is: "Wenn in beliebiger Zahl bewegliche Massenpunkte sich nur unter dem Einflusse solcher Kräfte be-

In this first statement of the principle, Helmholtz had not explained what the phrase "same relative position" meant. He had also allowed that a system might be considered to be in the same state with reference to a fixed centre; however, in the reply to Clausius, Helmholtz eliminates the passages in italics without comment. And, as I mentioned previously, he adds a further specification of what is meant by "being in the same relative position":

> Moving points have the same position relative to one another whenever a coordinate-system can be constructed in which all of their coordinates receive the same respective values.[22]

The reasons for the retraction and expansion of the definition of *vis viva* conservation are evident enough: Clausius's objection to Helmholtz was that there was no good reason why the force acting between two points might not be "an arbitrary function of the coordinates."[23] Helmholtz responded that the forces envisaged by Clausius would depend on magnitudes that were not determined by the mass-points of the system alone. Thus his previous admission of "fixed" centres had to be eliminated, for such a notion permits directional and positional relations that are not internal to the system. Furthermore, the scope of the term "same relative position" had to be more precisely defined, in order to block all appeals to properties of absolute space. Once these refinements had been made, he could argue that Clausius's assumption that the potential about a single point in space might vary with direction rests on a confusion between mathematically and epistemologically legitimate properties. The rebuttal of Clausius hinges, Helmholtz emphasises, on the demand that we must "seek the grounds of real effects only in the relations of real things to one another."[24]

The theoretical import of this change is two-fold: first, the energy principle is now formulated exclusively in terms of the "internal" properties of a system; second, the question of what it means to construct an

---

wegen, die sie selbst gegeneinander ausüben, so ist die Summe der lebendigen Kräfte aller zusammengenommen zu allen Zeitpunkten dieselbe, in welchen alle Punkte dieselben relativen Lagen gegeneinander einnehmen, wie auch ihre Bahnen und Geschwindigkeiten in der Zwischenzeit gewesen sein mögen."

22 "Gleiche relative Lage zu einander haben bewegliche Punkte, so oft ein Coordinatensystem zu construiren ist, in welchem alle ihre Coordinaten beziehungsweise dieselben Werthe wiederbekommen." (Helmholtz 1854), p. 83.

23 (Helmholtz 1854), p. 84.

24 (Helmholtz 1854), p. 84.

empirical coordinate-system is raised, if only implicitly. Regarding the first point, Helmholtz now requires that the energetic state of a closed system must be definable without reference to any external frame. Forces that violate this requirement, in that they would characterise systems for which Helmholtz's principle does not hold, are to be dismissed as mathematically possible, but epistemologically bankrupt. The assumption of such forces cannot "be applied [*übertragen*] to physical reality."[25] Even in his 1882 comments on the *Conservation*, where he admits that the principle of decomposition is an empirical proposition, Helmholtz continues to insist on this point when discussing the state of electromagnetism. Those theories that would make forces depend on absolute space, although they cannot be ruled out on purely logical grounds, are still the option of last resort.

The second point to observe is the modified status of geometrical relations entailed by this analysis. In order to rebut Clausius, Helmholtz is led to deny that directions and magnitudes which are not determined by "the real relations of things to one another" may be introduced into theoretical definitions. But his own definitions continue to employ the notion of "constructable coordinate systems" that will allow us to say of a system that it is in the same state at two points in time. This puts the ball back in Helmholtz's court. If it is true, as he maintains, that a single point does not determine *directions* in its vicinity, in what sense can it be said that two points determine a distance? It may well be the case that they determine a single spatial magnitude, that is to say a single line element. But that sense of determination is not sufficient to do the work required by Helmholtz's definitions: the magnitude they determine must be congruent with another magnitude determined by those same points *at a second point in time.* And Helmholtz, in contrast to Kant, is quite aware that there is a problem lurking here—that our concept of congruence cannot be defined as the "complete similarity and identity" of two spatial magnitudes, to repeat Kant's definition in the Phoronomy, but that every claim concerning the congruence of two spatial magnitudes contains an implicit reference to motion.

In an early memoir reproduced by Königsberger, Helmholtz had already identified this point explicitly, even if he failed to respond to it adequately. Here, Helmholtz distinguishes between what he calls "mathematical bodies" and "material bodies," suggesting that the first are like rigid bodies surrounding or containing the latter. Spatial magnitudes

25 (Helmholtz 1854), p. 84.

may therefore be conceived as "continuous rigid systems" in which the relative determinations of the system (by which Helmholtz means the relative distances and orientations of its points) are unchanged. He then proceeds to define congruence as possible superimposition:

> Rigid systems are congruent when the one can be moved onto the other in such a way that each point of the one coincides with a point of the other. Pairs of equally distant points are congruent. ... Motion must belong to matter quite aside from its special forces; but then the only remaining characteristic of a determinate piece of matter is the space in which it is enclosed; but since it is robbed of this characteristic as well by motion, we can only speak of its identity if we can intuit the transition from the one space to the other, i.e., motion must be continuous in space.[26]

In other words, Helmholtz explicitly problematises the notion of geometrical determination before he begins work on the *Conservation*, thus well before he appeals to the notion of a possible coordinate system in the reply to Clausius. In this manuscript, as in the much later papers on geometry, he considers the comparison of spatially distinct parts to be an ineliminable part of the *concept* of congruence.[27] Systems, whether "mathematical" or "material," are congruent if and only if they can be superimposed. Kant would so far be in agreement. But Helmholtz discerns a further difficulty: the motion of either kind of body involves a change in place, and this calls into question the identity of the system after the motion with that before the motion. He does not at this point provide a satisfactory analysis of what is involved in the concept of comparison itself, reverting instead to the supposition that we can intuit the formal identity of an object in motion, and thus that we can assure ourselves of the invariance of the system (namely, of a ruler-like "mathematical body") by appealing to continuity.

---

26 (Königsberger 1903), p. 185.

27 Again, compare Kant on the importance of coincidence (*Deckung*) in *Prolegomena* §12, 4.284: "All proofs of the complete equality of two given figures (since one can be placed in all its parts in the place of the other) finally amount to this: that they coincide with each other [*daß sie einander decken*]; which is obviously nothing other than a synthetic proposition resting on immediate intuition." Kant's immediate point is that the proposition does not depend on purely conceptual relations; however, he does not problematise here or elsewhere the sense in which two spatially *distinct* figures could be said to coincide. And, as we know from the Phoronomy, when he is confronted with physical laws that require the equality of distinct motions, he responds by insisting that they must be made strictly identical if geometrical propositions are to apply to them.

These early writings demonstrate that Helmholtz is already aware by the time of his reply to Clausius that the assertion that a single material system is in the same state at two distinct points in time cannot simply mean that the system occupies "equal" portions of space. Furthermore, if the system in question has undergone changes in the relative positions of its masses in the intervening period, then we cannot appeal to its continuous identity in order to intuit that they are now in the initial configuration. Since Helmholtz insists that such a statement must involve only "real" relations, we can infer that the possibility of constructing a coordinate system to which Helmholtz refers must at the very least involve the possibility of determining congruence relations by comparison with rigid "mathematical bodies" of the sort he referred to in the earlier manuscripts. In sum, since Helmholtz denies that directions can be employed in physical theories as if they were independent properties of space, he cannot himself assume that congruence relations can be so employed. This injunction against absolute directions is not, as Helmholtz emphasises, a "logical" one; rather it derives from a distinction between those coordinate systems which one "draws on paper"[28] and those which are determined by "real things." The meaning of the principle of energy conservation is that the energetic state of a system is the same whenever it is in the same state. But if we are to avoid covertly appealing to coordinate systems drawn on paper, we must admit that this statement assumes the existence of material bodies independent of the system in question that we can use to compare the system in its two states.

(d) Conclusion

Thus Helmholtz's position in his reply to Clausius is highly unstable, and this for several reasons. First of all, the principle of decomposition does not have the *a priori* status that Helmholtz implicitly accorded it, as he admits the moment he identifies its role. But since he needs it in order to reduce complex systems to two-point systems, he must also admit that non-central forces are not impossible on constitutive grounds. Second, the appeal to non-arbitrary coordinate systems does more work than Helmholtz could have wanted. According to Helmholtz, I cannot appeal to properties of space itself in saying that a system is in the same configuration at two different times. But according to his own ana-

28 (Helmholtz 1854), p. 84.

lysis in the Königsberger memoir on fundamental physical concepts, I also cannot appeal to the spatial determinations provided by the system itself, for these have, by hypothesis, changed in the intervening period. Thus Helmholtz is already committed to the existence of an empirically given coordinate system used to define those sets of a system's states that qualify as congruent. If he doesn't assume such a coordinate system, he has no empirically given magnitudes to ground his definition. Unfortunately, if he does admit one, he runs the risk that his opponent Clausius can employ it as well. For if I have enough measuring instruments at my disposal to determine the distances and angles that characterise the state of a complex system, why can't I use them to define asymmetries of the sort invoked by Clausius? The state of the system is no longer characterisable *solely* by means of internal relations. But Helmholtz cannot afford to relinquish that position, for to do so would mean to threaten the entire basis of his transcendental argumentation.

What Helmholtz needs is an argument proving the following: (1) statements concerning the relative positions of points in a material system are always statements concerning the (possible) coincidence of these points with other systems of points (the "mathematical bodies" he had discussed in the memoir); (2) the congruence relations determined by these mathematical bodies satisfy, or indeed entail, the axioms of (Euclidean) geometry. If (1) is true, then one is justified in rejecting any physical theory that appeals to intrinsic magnitudinal and directional properties of space on the grounds of empirical indeterminacy. In other words (1) amounts to saying that such relations are *necessary* to the empirical sciences. Conversely, (2) is needed in order to ensure that the resulting system of measurement satisfies all demands that physics places on its elementary magnitudes. In other words, (2) amounts to saying that such relations are *sufficient* to the requirements of the empirical sciences. These are peculiar demands, effectively requiring that one prove the validity of Euclidean geometry by deriving it from transcendental conditions on the empirical comparison of arbitrary spatial magnitudes. Yet this it is just what Helmholtz tried to do in his first two papers on geometry from 1868.[29]

Although they contain the bulk of his mathematical arguments, these papers are generally read as an abortive attempt at the arguments for the empirical status of geometry presented in the papers following them. But the philosophical argument is unmistakably a transcendental one. In

29 (Helmholtz 1868a, 1868b)

these papers, Helmholtz argues there that since physics presupposes our ability to equate spatial magnitudes, it also assumes that we make comparisons of the distances determined by pairs of points by transporting measuring instruments between them. Helmholtz then uses analytical methods to deduce the metrical characteristics of those manifolds in which the requisite measurement operations can be carried out. If such comparisons can be made regardless of the location and orientation of the points, this entails that there be rigid bodies that can be freely transported and rotated, and therefore, he concludes, that the possible manifolds thus singled out have a constant curvature. Adding to these assumptions the demand that space be unbounded, he concludes that only a space with a Euclidean geometry is compatible with the demands of physical measurement.[30]

Had he been right to conclude that Euclidean geometry was entailed by his measurement postulates, Helmholtz would have established that the basic magnitudes referred to in physical theories were empirical in the sense that Kant had when he called the propositions of the metaphysics of nature "pure empirical." They would refer to relations of coincidence among pairs of ideal *material* points. The supposition that geometry was true would then be equivalent to supposing that certain sets of relations among such ideal points in fact obtained. This would not be a constitutive *a priori* truth, but a regulative one: if it didn't hold, then certain kinds of physical descriptions would not be possible. Furthermore, the negative arguments that appealed to the indeterminacy of absolute spatial relationships would remain untouched. Helmholtz could still have maintained, against Clausius and others, that a theory which required reference to purely mathematical magnitudes was invalid. Here again, it would be so not because such theories were logically impossible, in that they would describe states of affairs which were unimaginable. Rather, they too would violate the regulative demand of the complete comprehensibility of nature.

Because Helmholtz formulated his later papers on geometry as empiricist arguments directed against dogmatic Kantians, and because his later readers have considered these writings in isolation from his research programme in physics itself, they have also downplayed the strongly transcendental reasoning that motivated the first two papers on geometry. On

30 This claim was, however, quickly corrected in an addendum from 1869 to the first of these papers, in which he admitted that pseudo-spherical spaces are also compatible with his demands. (Helmholtz 1868a), p. 617.

my reading, this aspect of the papers is not a philosophical atavism, but is directly related to the line of transcendental argument that we find in the *Conservation.* Indeed, since Helmholtz wrote these first two papers on geometry at the time when he turned his attention back to electrodynamics, in order to reopen the quarrel with competing electrodynamic theories, it is at least possible that his intent was to pick up his defence against Clausius and Lipschitz at the point he had left off in 1854. In Chapters 5 and 6, we shall consider in greater detail the relation between these later geometrical arguments and the research programme of the *Conservation.* Before doing so, however, we need to consider the sources of the specific method that Helmholtz applied to solve the problem of space. This was to *operationalise* metrical relations. Helmholtz's source here was an unexpected one, deriving from his research on colour-perception. In the following chapter, we shall see how the intervention of the mathematician Hermann Graßmann in that research programme afforded Helmholtz the methods he needed to render geometry a transcendental science with an empirical content. These methods allowed him to dissociate metrical characteristics of space from topological ones, and thus to distinguish rigorously between spatial magnitudes which are arithmetised and those which are not.

4. Colour-theory and Manifolds[1]

In the last two chapters, we have identified two thirds of the scientific and philosophical background to Helmholtz's philosophy of geometry. By considering the strategy that Helmholtz derived from Kant's *Metaphysical Foundations of Natural Science*, we saw how his attempts to invalidate non-central forces on transcendental grounds led him to call into question what he referred to as "purely mathematical" spatial relations. In Chapter 5, when we finally address the various alternatives considered by Helmholtz to such a "mathematical" conception, we will be able to characterise his position as a modification of Kant's theory of spatial magnitudes. In essence, Helmholtz will argue that we should collapse geometry and phoronomy, and thereby introduce a single, operationally defined concept of length. In the early papers on geometry from 1868, this operational definition is given in terms of rigid body displacement: two regions of space are congruent when it is possible to displace a rigid body from one to the other; two objects are of equal length if they coincide whatever their situation. It is perfectly possible, Helmholtz emphasises, that there be no bodies whose displacements could allow us to establish these equivalence classes, and thus geometrical propositions have an empirical basis. In the later papers, Helmholtz quietly retrenches, replacing his rigid-body definition with a properly phoronomic one in his last paper, from 1878. He introduces the notion of "physically equivalent magnitudes," which in the simplest case are distances traversed by inertially moving bodies in the same period of time. The science that describes the relations among such magnitudes, which he calls "physical geometry," should replace "pure intuitive geometry." For the equivalence relations among the spatial magnitudes treated by the latter science are ones that we merely intuit—they lack any connection to actual physical processes, and indeed if they were found to have such a connection, this could be no more than a lucky accident.

1 This chapter was first published in a somewhat different form as: Hyder, David. 2001. Physiological Optics and Physical Geometry. *Science in Context* 14 (3):419–456. Copyright Cambridge University Press. Reprinted with permission.

Throughout these papers, Helmholtz holds to the claim that the purpose of geometry is exhausted by its role in the formulation of general physical laws. Neither the "metrogenic apriorism"[2] of the first two papers, nor the phoronomic definition of physically equivalent magnitudes in the last one make any sense outside of this context. Because we are driven to codify our experience in terms of general laws, and because empty space is undifferentiated, we require a system of measures that will allow us to describe motions by means of equations. Because empty space is unobservable, this system of measures cannot be grounded in properties of space itself, that is to say, it cannot be conceived as a description of the magnitudinal structure of absolute space. Physics needs its coordinate-systems, but the latter cannot be "purely mathematical," as Helmholtz admonished Clausius. Thus there must be some alternative explanation of what these coordinate systems are.

As I suggested in the Introduction, this decoupling of coordinate-systems, and of the metrical relations they define, from the purely topological properties of space, such as dimensionality and continuity, is the defining mathematical characteristic of Helmholtz's critique. It is coupled to the epistemological claim that, if a coordinate-system and a metric can be defined for space, then this is the result of operational definitions that are chosen to best serve the regulative aims of physics. On its own, space has no metrical structure. Indeed, one might observe parenthetically that Helmholtz's unwillingness to consider metrics of non-constant curvature, with which he was of course familiar from Riemann's work, is explicable in the light of this last conviction alone. Because such metrics ascribe a non-uniform structure to space, they violate the basic Kantian tenet that space is a completely neutral form of possible experience.

In this chapter, I shall add the third source of this view, namely the research programme pursued by Helmholtz and others between 1852 and 1866 in colorimetry. At its inception, this research had little to do with the theory of spatial magnitudes. Helmholtz's intention was to investigate Newton's theory of colour-mixture, and no doubt to defend it against rival theories that had emerged in the meantime. But the intervention of the mathematician Hermann Graßmann in the debate, as well as the exceptional contributions of the young James Clerk Maxwell, quickly transformed the nascent topic of colorimetry into a field-test of Graßmann's vector calculus, which the latter interpreted as a theory of "additive magnitudes." The colour-space was no longer interesting merely

2 The term is due to (Carrier 1994).

for sense-physiologists, for in Graßmann's hands, it became the single alternative example of an intuitively given manifold. In the course of this research, Helmholtz encountered theoretical and practical difficulties concerning the sense in which one could speak of "distances" within the colour-space, problems that were analogous to those which he, and Riemann, eventually came to identify in the theory of geometry. As the discussion of the later chapters will make clear, Helmholtz drew on these lessons throughout his later work.

This chapter draws on a number of recent studies on Helmholtz's work in colour theory, but differs from them in an important respect.[3] Since previous authors have been concerned with physiological optics *per se*, they have understandably identified only in passing, if at all, those aspects of that work which are related to geometry. Most important of these is the emergence of the distinction between measuring *sensations*, and measuring their causes. A preliminary statement of this difficulty may serve to keep the main line of argument clear. The result of the research programme described in the following was the "trivariance" or three-receptor theory of colour perception, which says that colour impressions are produced by the actions of three fundamental colour receptors. These receptors were assumed by both Maxwell, and later Helmholtz, to respond differently to light of varying wavelengths. The reason that colours can be organised in a plane, or space, is that one can regard these receptors as the physical grounds of three independent variables defining a continuum. But the three-dimensionality of the space can be captured by other variables, such as colour-tone, brightness, and hue. Just as physical space can be *described* by any set of three linearly independent vectors, the same holds true of the colour-space. But unlike physical space, the colour-space has an objective ground,[4] for one set of such vectors—those physically realised in the receptors—reflects the unique cause of the space's existence. Furthermore, the responses of these receptors to physical light presumably give the colour-space a uniquely correct metrical structure.

A central difficulty that emerged in Helmholtz's and Maxwell's work after 1855, however, revolved around the following difficulty: they were convinced that the above account was correct, but they had no way of

3 (Kremer 1993; Sherman 1981; Turner 1994)

4 The point is merely that we do not believe that there is a single "correct" set of basis vectors that describe physical space. I am not, quite obviously, going to speculate about the causal basis of physical space in this book.

identifying the primitive receptors, let alone specifying exactly their response curves. Later research (which shall not be treated here) centred on developing methods to do this. But both in 1855, and in the first edition of Helmholtz's landmark *Handbuch der physiologischen Optik*[5] in 1860, the claim that one had identified the "empirical" structure of the colour-space was untenable. One may make this point simply as follows. Assume that you accept the trivariance theorem. You also accept that the excitation of the three receptors in response to light of a given wavelength can be quantified, because it is a physical process. Does it follow that the subjective impressions of the three fundamental colour-sensations can be compared quantitatively in the same way?[6] There is no inherent quantitative relation between pure impressions of red and green. Thus any mapping of subjective colour-impressions onto a spatial structure will simply ascribe them metrical relations based on our intuitions of which are "equally bright" and which are "close to" of "far from" each other. But these subjective relations only acquire an objective meaning when they are tied to objective properties of light by means of a measurement procedure. In his last paper on geometry, Helmholtz criticised the notion that we could have intuitions of equal spatial magnitudes (or that, if we did, they had any privileged significance) on just these grounds: it is only the connection between spatial magnitudes, which have been selected as equivalent by means of a measurement procedure, and physical processes that gives the former objective meaning; even if we did have immediate intuitions of such equal distances, we would have no reason to accord them a special significance.

In order to see how these difficulties emerged, I discuss four publications in detail: Helmholtz's 1852 "Über die Theorie der zusammengesetzten Farben" (On the Theory of Composite Colours); Graßmann's 1853 "Zur Theorie der Farbenmischung" (On the Theory of Colour Mixture); Maxwell's 1855 "Experiments on Colour"; and Helmholtz's 1855 "Über die Zusammensetzung von Spectralfarben" (On the Composition of Spectral Colours).[7] Although my subject is Helmholtz, the work

5 (Helmholtz 1867). The first volume appeared in 1860.

6 The problem of specifying a procedure for quantifying sensation led to an entire research programme, founded by Fechner, in "psychophysical" measurement. Whatever the merits of that later work, in the 1850's the results were not yet in. On the contrary, the problems encountered there were instrumental in bringing the latter field into being.

7 (Helmholtz 1852a, 1855; Graßmann 1853; Maxwell 1855)

he drew on in this period was not his alone: a key role was played by the mathematician Hermann Graßmann, as well as by the young James Clerk Maxwell. It was only considerably later that Helmholtz adequately acknowledged Graßmann's influence on his work, mainly in the second edition of the *Handbook of Physiological Optics*, as well as in his late work on measurement, *Zählen und Messen (Counting and Measuring)*. Maxwell was never given his full due. In any event, there can be little doubt that the formulation of the problems that, in my view, were then transferred by Helmholtz to geometry proper, was essentially determined by the two other men. Since all four papers were attempts at analysing and developing the mixture-table, or colour-wheel, that Newton had described in his *Opticks*, I begin by outlining Newton's theory, and then discuss each of the four papers in chronological sequence. I will conclude by tracing out the connections between the colour-space (which is what it had become by the end of this period) and Helmholtz's geometrical papers. These connections will then be treated in greater detail in the two following chapters.

(a) Newton's Barycentric Colour-wheel

The "barycentric" colour-wheel was presented in Newton's *Opticks* as a calculating device, whose function was to permit one, "In a mixture of Primary Colours, the Quantity and Quality of each being given, to know the Colour of the Compound."[8] The colours of the spectrum were arrayed around the circumference of the wheel (Fig. 1) in arcs of seemingly arbitrary length at each of whose middle-points a load-point (p, q, r, …) was inscribed.

Newton explained that in order to locate the colour resulting from mixing two or more spectral colours, one had simply to locate that point in the wheel which lay at the "Centre of Gravity"[9] of the several colours. That is, one imagined a weight placed at the appropriate load-point of each of the colours in the mixture, and determined the point z in the circle that would lie at their centre of gravity. The weights of the components were to be determined by "the Number of Rays" of each spectral colour entering into the mixture. One could then find the colour of the mixture by extending a line from the centre $\boldsymbol{O}$ through

8 (Newton 1931), p. 154.
9 (Newton 1931), p. 155.

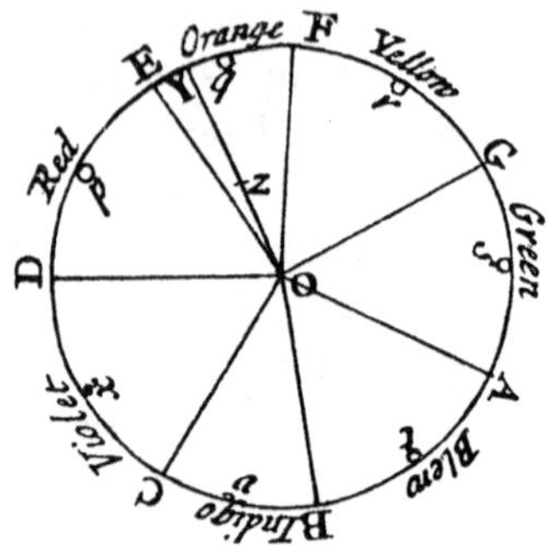

FIG. 11.

Fig. 1: Newton's Colour-wheel

z. The distance of *z* from ***O*** would "be proportional to the Fulness or Intenseness of the Colour, that is, to its distance from Whiteness."[10]

Newton did not think that the spectral colours arrayed around the circumference of his wheel were properties of light itself.[11] But he assigned a different status to the "Rays" evoking these colours: white and (presumably) mixed colours, which are always paler than their spectral counterparts, do not stand in a one-to-one relation with a kind of ray, as does the sensation of red to a "Rubrifick" ray,[12] and this difference is encoded both in the form of the colour-wheel and in the accompanying directions for its use. By selecting points on the circumference, one determines the locations of points on the interior. But no line connecting interior points ever intersects the exterior, and each chord passes through only interior points, which geometrical relations reflect the fact that mixed colours cannot yield pure ones, and that no pure colours mix to form another one, respectively.

Newton did not explain how he had arrived at this exact form for the wheel, nor what he meant by "the Number of Rays" that each colour in the mixture was to have. He also did not explain what, if anything, the

10 (Newton 1931), pp. 155–56.

11 "Colours … in the rays are nothing but their Dispositions to propagate this or that Motion into the sensorium, and in the Sensorium they are sensations of those Motions under the Forms of Colours." (Newton 1931), p. 125.

12 "The Rays in that mixture do not lose or alter their several colorific qualities, but by all their various Kinds of Actions mix'd in the Sensorium, beget a sensation of a middling Colour…." (Newton 1931), p. 158.

wheel might represent, that is, whether it was anything beyond a device that allowed one to arrive at an answer to a particular question.

(b) Helmholtz's 1852 "Über die Theorie der zusammengesetzten Farben"[13]

Newton's theory distinguished between pure and mixed colours, but it did not have an equivalent to the primary colours which formed the basis of colour-tables in the industrial arts, and which had played a central role in the theories of Young and Brewster.[14] Helmholtz divided the primary-colour theories into three groups, "explicitly fram[ing] his investigation as a test of these agnostic (painterly), physical (Brewster), and physiological (Young) views of the primaries."[15] In Helmholtz's words:

> 1) either that the primary colours are those out of which all possible others may be composed;
> 2) or, as with Mayer and Brewster, that the primary colours correspond to three objective kinds of light;
> 3) or that they correspond to three distinct primary sensations of the optic nerve fibres, out of which the other colours sensations are composed, as with Thomas Young.[16]

The question Helmholtz set out to answer was, in other words, the following: Where does colour-mixture take place? These various theories made room for a wide variety of experimental results, and one may take it that Helmholtz's aim was to decide among them as best he could, as opposed to discrediting them all. But even this reading may credit Helmholtz's rhetoric too much, for he, in the opening paragraph of the paper, makes quite clear which option he prefers:

> Light rays of various wavelengths and colours differ in their physiological action from tones of various vibration periods and musical pitch essentially in that each pair of rays, acting simultaneously on the same nerve fibre, calls forth a simple sensation, in which even the most practised sense-organ cannot distinguish the individual combined elements.... This unification of the impression of two distinct colours in a single new colour impression is ob-

13 (Helmholtz 1852a)

14 Accounts of the theories of Brewster and Young can be found in (Kremer 1993; Sherman 1981).

15 (Kremer 1993), p. 228.

16 (Helmholtz 1852a), p. 7.

> viously a purely physiological phenomenon, and depends only on the characteristic reactions of the optic nerve.[17]

Whereas a practised ear could distinguish the component simple tones in a complex chord, the same was not true of complex colours; and whatever the number of fundamental colours, the bulk of those we sense are the results of *physiological* mixture. As Newton before him, Helmholtz took spectral light-rays to be simultaneously of "differing wavelength and colour."[18] For Newton, fundamental light-rays of a given refrangeability and colour were such that—whatever their phenomenal appearance—they could not be decomposed by physical means, for instance prismatically. A primary-colour theory suggests, however, that some of the colours around Newton's wheel are themselves "mixed." And so the questions arose: What is mixed? And how? Brewster's claim had been that the light itself was mixed, and that prismatic decomposition was not the last word: he claimed to have decomposed prismatically "pure" light with filters. Helmholtz rejoined in a simultaneous article against Brewster that he had been unable to repeat these experiments.[19] Such a separation could not be achieved if the spectral light had been adequately purified. Of course this could not rule out the possibility that some third method (neither refractive nor filtrative) would not reveal, say, spectral yellow light to be mixed. But it would have to be shown that the lights in question were indeed pure in respect of their wavelengths if such a positive conclusion concerning their physical composition were to be drawn.[20] It followed that, if anything could meaningfully be said to be mixed, it would have to be the sensations. Nonetheless the variables subject to experimental control could only be the wavelengths and the intensities of pure spectral light, for there was no way of operating on the nerve-fibres directly.

Although he had designed an apparatus which afforded him control of both light intensity and colour, Helmholtz was able to replicate Newton's results only partially in 1852: some chromatometric aspects of the colour-wheel were maintained (red and yellow were shown to yield an orange which was barely distinguishable from spectral orange), but the

17 (Helmholtz 1852a), p. 3.

18 "Es bleibt also der von Newton angenommene Zusammenhang der Schwingungsdauer oder Brechbarkeit mit der Farbe unverändert bestehen." (Helmholtz 1852c), p. 600.

19 (Helmholtz 1852b).

20 (Helmholtz 1852b), p. 44.

overall topology of the wheel, that is to say the generality of Newton's barycentric equations, could not be maintained. For Helmholtz could not find more than a few of the complementary pairs (pairs of spectral lights which mixed to yield white) that Newton's wheel predicted. Contrary to his claims in his next paper on the subject, the problem was not that he had been misled by "peculiar physiological properties of the human eye."[21] Rather, the method he had used had not allowed him to produce enough spectral light, nor indeed to adequately control its intensity. Helmholtz mixed his spectral lights by superimposing two spectra, and adjusting their widths. The quantities of lights entering into the mixtures could be only roughly adjusted, and certainly could not be measured. But this difficulty only concealed a further one, which was to become critically important later: in what sense could one's attaching measures, however imprecise, to the *spectral lights* be seen as equivalent to measuring *phenomenal colours?*

Response to Helmholtz's 1852 paper was critical, which is not surprising, when one considers his largely negative conclusions. Brewster's theory, Young's theory, indeed Newton's theory were all wrong; and although it was possible to mix a large number of pigment-colours from only a small number of pigments, the same did not hold for the lights which conveyed their colours to the eye. Helmholtz himself could not have been satisfied with his results, which contradicted common sense. After all, according to his results, even pigments mixed more predictably than light, and they did so from a smaller number of primaries. But it is easy to measure the quantity of a pigment, whereas measuring the wavelength and quantity of spectral light is not. So in the following years, armed with a far more sophisticated apparatus, Helmholtz returned to the fray.[22] But in the interim, the mathematical basis for his research had been radically changed.

(c) Graßmann's 1853 "Zur Theorie der Farbenmischung"

Probably the most important stimulus to this further work was a paper published by Hermann Graßmann in 1853 on the "Theory of Colour-mixture." Today recognised as the founder of vector analysis, Graßmann was also an accomplished philologist, publishing a translation and dic-

21 (Helmholtz 1855), p. 46.
22 *Cf.* (Kremer 1993), p. 232.

tionary of the Rig-Veda, as well as on phonology.[23] Graßmann's paper applied the calculus of his neglected 1844 *Lineale Ausdehnungslehre*[24] to the structure of Newton's colour-wheel, introducing four principles of colour-mixture which were used as formal axioms to derive the geometrical properties of the wheel. The wheel became, on this analysis, a two-dimensional cut through a three-dimensional colour-space, whose dimensions were, in modern terminology, the hue, brightness and saturation of the various colours. But this description does not do justice to Graßmann's insight, for the most remarkable feature of the *Ausdehnungslehre* was that it described spaces as the products of *operations* with fundamental *additive magnitudes.*[25]

According to Graßmann's analysis, each colour impression was completely determined by three magnitudes, representing the colour-tone (the hue), its quantity (brightness), and the quantity or intensity of admixed white (the saturation).[26] Spectral colours retained the special role they had in Newton's analysis: only they had a direct relationship to wavelength, and all other colours were composites of them. This property had an exact mathematical correlate in Graßmann's axiomatised system: the value of the third magnitude (the saturation) of the pure spectral colours was always zero, i. e. the intensity of a spectral colour was represented entirely by the second magnitude, the brightness of that colour, whereas the intensity of a mixture was distributed between the magnitudes for brightness and saturation.[27] A mixture of two (or more) spectral colours is located, as with Newton, at a point on the interior of the circle which represents the "centre of gravity" of the colours in question. But on Graßmann's wheel, the distances of that point from (a) the centre and (b) the periphery, represent its brightness and "quantity of admixed white" (its saturation), respectively. So Graßmann used his two-dimensional surface to represent three distinct quantities. He summarised these relations in a single fundamental proposition:

23 In the 1870's, Graßmann published a paper paralleling the one discussed here, in which he attempted a planar arrangement of phonemes by means of his *Ausdehnungslehre.* The *Ausdehnungslehre* itself was modelled on Leibniz's universal characteristic.

24 (Graßmann 1878)

25 *Cf.* (Beutelspacher 1996), p. 5.

26 (Graßmann 1853), p. 70.

27 This relation was secured by his fourth principle, on which more in the following.

> From this it follows that, if one conceives the entire mass as unified in the centre of gravity, in which case one calls the centre of gravity loaded with such a weight the *geometric sum* of the individual points loaded with their weights, then every light-impression can be represented exactly, in accordance with its three moments, by a given point loaded with a given weight. The direction from the centre in which this point lies, or, just as well, the point at which this direction intersects the circumference, represents the colour-tone. The weight of the point represents its total intensity. The product of this intensity and the distance from the centre represents the intensity of the colour, and the product of the intensity with the distance from the circumference represents the quantity of admixed white.[28]

For Graßmann, the location of a point within the circle specifies the colour-tone and the *ratio* of brightness to saturation, and the absolute intensity of the mixture in question is given by a distinct value, represented as a weight. The location of a given mixed colour would be determined by means of the weighted vector sum of the colours entering in the mixture (a "geometric sum" in the terminology of the *Ausdehnungslehre*). The mixture-law can be described by means of a two-dimensional representation only because Graßmann's fourth law—that the intensity of a mixture (the combined brightness and saturation, or whiteness) equals the sum of the intensities of its components—made those values linearly dependent.[29] Although he presented this analysis as a pious defence of Newton's doctrine, Graßmann achieved with it a level of generality that was at best implicit in Newton's own account. The truth of his axioms conceded, Graßmann proved formally that Helmholtz's results had to be wrong. This proof is of particular interest because it depended only on his axiom of continuity (each colour would change continuously as its brightness, spectral colour or saturation was changed) and on the assumption that the spectral colours could be arranged in a closed loop about the wheel within which the point representing white was enclosed. In other words, Graßmann argued from topological characteristics of the col-

28 (Graßmann 1853), p. 82.

29 "daß die gesamte Lichtintensität der Mischung die Summe sei aus den Intensitäten der gemischten Lichter. Hierbei verstehe ich unter der gesamten Lichtintensität die Summe aus der Intensität der Farbe, wie ich sie oben festgestellt habe, und aus der Intensität des beigemischten Weiß …" (Graßmann 1853), p. 82. That is to say: the mixture law is composed of a function mapping four variables (the colour-tones and brightnesses of the components) onto two (the colour-tone and brightness of the mixture) and a second function mapping two independent variables (the intensities of the components) onto one (the intensity of the mixture). Newton's wheel did not clearly distinguish these relationships.

our-plane that there had to be more complementaries than Helmholtz had been able to find.

Two points stand out: (1) Spectral colours are distinguished within the mathematical representation only in their having no saturation or whiteness, that is, only in their third coordinate being zero; (2) Graßmann's third law, which states that the actual composition of a given colour is irrelevant to its effects in a given mixture (mathematically, that all sets of vectors summing to a given vector are equivalent as regards the contribution of that vector to further sums) suggests that the choice of those colours which serve as a basis for a given portion of the colour-wheel is—mathematically—irrelevant to the relations of colour addition which hold on the wheel. Thus the additive properties holding within the wheel can effectively be dissociated from any one set of vectors spanning it (that was indeed the strength of the vector-calculus Graßmann had developed). Of course each spectral colour was still assumed to be a physically unmixed primary, even though the space of all colour-impressions was definable by means of only three variable quantities, as was the one operation—mixture—on this space. For the fact that the vectors added only positively meant that none of the spectral colours could be synthesised: any addition of colours of distinct tones would decrease the saturation, that is increase the whiteness, of the mix. Thus it followed that only the full complement of spectral colours sufficed to span the colour plane under the single additive operation which Graßmann allowed.

We can illustrate the exceptional role of the primary colours by means of a naive question: What lies outside Newton's and Graßmann's colour-wheel? On their intended interpretation, the answer is: Nothing, for the points on the circumference represent colours of maximal tone-intensity and minimal whiteness; these points lie at the limit of possible perception. At the same time, however, the planar representation and the mathematics that describe it do permit a hypothetical answer: the points outside the colour-wheel would represent colours of still greater tone-intensity, with negative whiteness. These are what we would today call supersaturated colours. Such colours could provide a basis for a space greater than that of the colour-wheel—a space within which the spectral colours themselves could be additively synthesised. As we shall see in a moment, this is exactly the interpretation proposed by Maxwell when he revived Young's theory, thereby articulating the theory which Helmholtz eventually adopted. So it is worth inquiring what speaks (or spoke) against this move.

Newton, as we saw, ascribed his spectral colours a completely different status from those mixed out of them. They lie on the perimeter of the colour-wheel, and they stand for physically unmixed lights. At the same time, they are the colours that mix to yield those on the interior of the wheel. In other words, the spectral colours *denote* spectral lights directly, and these lights are simple data out of which complex ones are generated. In contrast, the colours *within* Graßmann's and Newton's wheels have neither of these properties. They are, we might say, equivocal: each of them may be the product of a number of different mixtures; furthermore such mixtures may involve further (mixed) colours that may themselves be equivocal in just this sense. None of these authors consider that the primitive elements of sensation-mixtures might have *no* unique correlate in what Helmholtz called the "objective" properties of light. Even if Helmholtz was prepared in 1852 to consider the question of whether, say, the sensation of pure spectral yellow is "mixed," he still assumes that the fundamental elements of that mixture will themselves be sensations that denote pure spectral lights. And so even Graßmann, whose mathematical analysis opens up the possibility of abandoning this assumption, still argues, at the end of his paper, that the first question to settle is how the "homogeneous colours" are in fact laid out around the circumference on Newton's wheel.[30] Helmholtz, as we shall see in the next section, did not quite agree with this: in his 1855 paper on colour-mixture, he assumes that determining the *shape* of Newton's wheel is just as important. He does not, however, doubt that this can be settled "objectively," namely by determining the relationships holding between what he too calls "homogeneous colours."

(d) Helmholtz's 1855 "Über die Zusammensetzung von Spectralfarben"[31]

What Helmholtz needed to achieve this goal was what the French physicist Léon Foucault graciously afforded him: "a simple method for combining in any proportion all kinds of simple rays, determined by their position in the spectrum, with the same rigour and ease as when mixing ma-

30 (Graßmann 1853), pp. 83–84.
31 (Helmholtz 1855)

terial colours."[32] Foucault's method allowed one to project two pure spectral colours onto a large surface, while screening out other light sources. More importantly, however, it allowed precise control of the quantities of the spectral lights entering into the mixture. Helmholtz made a few improvements to Foucault's apparatus, and with this new set-up he was able to overcome the problems on which his earlier research had foundered, and which had let him to reject Newton's planar representation of colours.

As I explained above, Graßmann had argued that from his first two principles—that (1) all mixed colours can be represented as a mixture of a spectral colour with white, and that (2) a mixture changes continuously when one or the other of its components changes continuously—it followed that each spectral colour would have a complementary. This argument presupposed that the spectral colours themselves could be arranged in a continuous loop, i. e. the circumference of the colour-wheel.[33] Since his proof rested on assumptions of continuity and ordering only, it did not show *which* pairs of colours would be complementary, nor in what proportions such pairs would have to be combined. Helmholtz's improvement to Foucault's apparatus lay in his being able to control both the locations and the widths of the slits on his plate—thus the colour-tones and their intensities—continuously while the apparatus was in place. This enabled him to isolate many more complementary pairs than he had been able to do in 1853. These pairs were of particular interest precisely because they were assumed to lie on opposite sides of the colour-wheel, of which white was the colourless centre, thus their positions could define—empirically—its shape.

It was the *shape* of the curve, however, which lay at the root of Helmholtz's disagreement with Graßmann on the significance of the latter's

32 (Foucault 1853a) reprinted in (Foucault 1878), p. 52. German translation: (Foucault 1853b).

33 Graßmann's proof is too intricate to give in full. In essence, he proves that (1) Each colour A has a colour A' "opposite" it on the colour-wheel, in that colours on either side of A' produce distinctly coloured sets of mixtures when mixed with A in varying intensities; and, on the assumption that A has no complementary, this property of A' leads to a contradiction, for (2) If one considers two colours B and C infinitely close to A' on either side of it, it follows that (a) the sets of mixtures of B and C with A must differ only infinitesimally, although (b) A' was defined in (1) in such a way that the members of the two sets consist of distinct colour tones. Since (a) and (b) contradict one another, A' must produce white when mixed in some proportion to A. (Graßmann 1853), pp. 73–75.

fourth principle. That principle stipulated that the intensity of a mixture of lights be equal to the sum of the intensities of the lights composing it. But, Helmholtz observed,

> This fourth principle can be applied in three essentially different ways, depending on the method one uses to lay down the intensity. First, one could call the light-intensity of distinct colours equal when they appear equally bright to the eye.... Second, one could declare the determination of the measurement-units of differently coloured lights to be arbitrary, and take this fundamental principle to mean that there was a way of determining the units according to which the light-intensity of the mixture was always equal to the sum of the mixed lights. ... The third way is this: one says that one has found a method for determining the intensities of different coloured lights, and one demands that the fundamental principle be correct for this particular way of reckoning intensities as well. This is how Graßmann proceeded.[34]

In a detailed scholium to the second interpretation of Graßmann's principle given here, Helmholtz showed that the shape of the curve bounding the colour-field depends in large measure upon arbitrary assumptions. In choosing any three colours (mixed or unmixed) and postulating their locations and units of brightness, one would of necessity fix the location of every colour that might be composed out of them. By extension, one would specify the location of all further colours, including the spectral ones. Assume we have chosen *A*, *B*, and *C* as our basis colours, and have determined the proportions—expressed in terms of arbitrary units for the intensities of the three colours—in which they mix to yield *M* (Fig. 2). Then a colour *E* lying outside the triangle *ABC* (that is, which cannot be mixed out of them directly) can be located, and its unit of intensity defined, by finding a further colour $M_{,}$ lying inside of *ABC*, and which in turn is produced by mixing *M* with *E*. *E*'s location and brightness are functions of those of *M* and $M_{,}$, and their locations and brightnesses depend only on those stipulated for *A*, *B*, and *C*. Thus *E*'s location and brightness also depend only on the latter stipulations.

This method allows for a certain conventional latitude. Depending on the three colours chosen, the locations assigned them on the plane, and the methods for measuring their intensities, the other colours could be distributed in many different ways. Each colour-plane so determined could be mapped, by means of linear transformations, onto the others, while no one of them could be seen as *the* colour plane. In each case, the spectral colours, although still lying at the boundary of

34 (Helmholtz 1855), pp. 67–69.

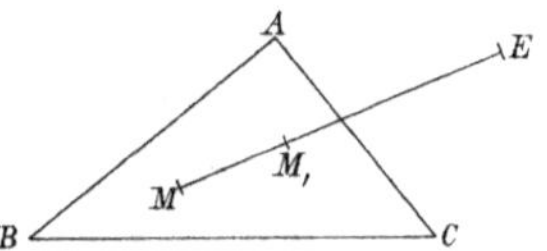

Fig. 2: Locating colours on the mixture plane.

the colour-plane, would also be arrayed in accordance with the choice of *A*, *B*, and *C*, and the units of their intensity. As I remarked above, the spectral colours were represented, in the mathematics of Graßmann's representation, no differently from the mixed ones (all were vectors, or *additive Größen*); however, Graßmann's method for constructing the wheel assumed that they were laid down first. Helmholtz here identified the possibility, although he did not make use of it, of reversing Graßmann's procedure, and thereby determining the locations of spectral colours on the basis of stipulated locations for mixed ones.

Thus according to the second interpretation of Graßmann's fourth principle, the shape of the colour-wheel, and the distribution of colours within it depended essentially on conventions of measurement: in modern terminology, all these colour planes had the same affine structure, and the various possible arrangements represented nothing more than various choices of basis colours and their units. But, contrary to what one might expect, Helmholtz did *not* conclude that each such plane was equally valid. For, on his analysis, the first and third interpretations of Graßmann's fourth principle did allow one to fix the shape and distribution, albeit on quite different assumptions: in the first case, one would say that two colours were equally bright when they *appeared* to be so, quite independently of their physical characteristics, and one would then derive the colour-plane implied by that stipulation; in the third case, one would stipulate the conventions of measurement and the form of the spectral curve which were to hold, while altering the arrangement of the colours to conform with these stipulations.

Helmholtz contended that Graßmann's interpretation of his fourth principle was too lax. Graßmann had defined the intensity of the spectral lights to be such that equal quantities of complementary lights mixed to form white light, from which it would follow analytically that all spectral colours would be equidistant from a white centre.[35] But his method could lead to unforeseen and unintuitive results. In fact, Graßmann's assump-

35 In fact, this requirement could be met for up to at most five spectral colours, so the resulting curve would not be perfectly circular.

tion that the spectral colours formed a loop could only be saved by admitting mixed colours to the circumference of the wheel, for there are no spectral colours "between" violet and red.[36] Perhaps more seriously, on Graßmann's approach, perceived or apparent brightness would in turn have to be sacrificed to that of brightness as defined in accordance with the above definition. And, finally, the arrangement[37] of the colours around the periphery of the plane would depend on these stipulations as well.[38] In Helmholtz's opinion, Graßmann's approach saved the mathematics at the expense of the phenomena. Such a colour-plane could serve, as it had for Newton, as a means of calculation, but its structure would correspond directly neither to the objective structure of physical light, nor to the subjectively experienced, intuitive relations holding among colour-sensations.

Helmholtz opted instead for the first interpretation of Graßmann's principle: define spectral colours as equally intense when they are perceived as being so, and let the mixed colours fall where they might on the plane. Projecting two complementary lights on his screen, he adjusted the width of the slits[39] until the mixture showed white. After measuring the width of the slit of the (apparently) brighter light, he placed a rod in the path of the lights and narrowed this slit until the two coloured shadows thrown by the rod appeared equally dark. He then measured the slit at its new setting. He set the ratio of the intensities needed to produce white to the ratio of the slit's widths in the two settings. He found that, green-yellow being apparently brighter than violet, the slit-width for green-yellow at which the two colours showed white was one-tenth

36 In (Graßmann 1877) he accepted this result, but felt his method basically untouched by this development.

37 Not their order, but their proportion of the circumference: certain ranges of colours would occupy more or less of the circumference.

38 He complained in 1878 that his demonstration "must have [contained] significant conceptual difficulties, since even Helmholtz did not fully understand it." (Graßmann 1877), p. 218. He, in turn, does not seem to have understood Helmholtz fully. As I discuss below, the dispute and misunderstandings derived from their quite different interpretations of the nature of the colour-space itself, that is, of what it in fact represents.

39 Helmholtz's apparatus worked by projecting a refracted spectrum on a plate with adjustable slits. The light passing through the slits was then focused on a screen, where one could then observe the mixed colour. By adjusting the placement and width of the slits, he could select different shades of spectral light and different intensities, respectively. Thus the width of the slit here corresponds to the quantity of light going into the mixture.

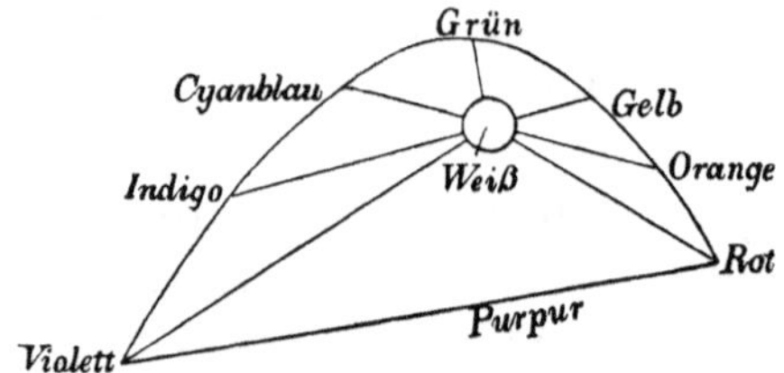

Fig. 3: Helmholtz's "objective" colour-wheel.

the width at which they threw equally dark shadows. Thus the ratio of the brightness (intensity) of spectral violet to spectral green-yellow was 1:10, and the "distance" of green-yellow from white would be one-tenth that of violet from white. Proceeding in this manner with the other three pairs of complementary colours he had found, he defined the relative weights of each pair, and supplemented these numerical relations with informal observations of mixtures that were not white (red with equally bright green was found to give a reddish orange, *etc.*).[40] He graphed these results in a flattened colour-wheel (Fig. 3). Since Helmholtz, in violation of Graßmann's expectation, had been unable to find a spectral purple, he filled in this portion of the spectrum with a line of purplish colours produced by the various mixtures of red with violet. They served as ersatz complementaries to green. He traced the remainder of the curve roughly in accordance with the weights determined in the slit-adjustment experiment.

(e) Maxwell's 1855 "Experiments on Colour"[41]

Helmholtz was not the only one to read Graßmann's paper. It is possible, given his discussion of the colour-space in the opening sections of his "Über die Hypothesen, welche der Geometrie zu Grunde liegen," that Riemann had done so as well.[42] More importantly, Maxwell drew on it in the research he presented to the Royal Society of Edinburgh in March 1855 in a paper entitled "Experiments on Colour." In this paper, Maxwell applied Graßmann's principles to construct a normal colour-plane, which he then compared with ones generated by red-blind subjects. He offered the differences between the two as evidence for a tri-

40 (Helmholtz 1855), p. 64.
41 (Maxwell 1855)
42 (Riemann 1854), p. 274.

chromatic theory, and thereby both explained the physical basis of Graßmann's three independent variables and opened the way for regarding the set of colours as a manifold in Riemann's sense of the term. The paper had three parts: first, Maxwell described in detail a method for situating colours on the plane determined by three arbitrarily chosen basis-colours; second, he advanced a somewhat improved version of Young's trichromatic theory, suggesting that every colour was the product of three elementary nervous processes; finally, he produced evidence for this claim by comparing the colour-planes of normal subjects with those of the colour-blind. So Maxwell committed himself first to a version of Young's trivariance theory, which Helmholtz had rejected in 1852 and simply did not mention in his 1855 paper.

Whatever the theoretical overlap between the two men's work, Maxwell had a decided advantage over Helmholtz in one respect: he used a low-tech colour-top (which he called a chromatic teetotum), instead of the high-tech light-mixing apparatus available to established researchers like Foucault and Helmholtz. Maxwell's teetotum allowed him to produce incomparably more—and in some respects more precise—data than Helmholtz had been able to do. The teetotum's operation was simple: one arranged coloured swatches on the surface of the wheel; one then observed the colour that resulted by spinning the top. The colours were perceived as a single fused colour, because the light impinging on the retina of the observer rapidly alternated as the top spun.

Let me briefly summarise Maxwell's method in one case, in order to see exactly how his method was applied. By altering the areas occupied on the top by different colour swatches, Maxwell could generate accurate numerical values for the quantity of each colour going into the mixture. He did not compare *coloured* mixtures to one another, but to the central region of his top, on which separate black and white papers were set to produce a distinct shade of white/grey. If, say, three colours could be arrayed on the periphery of the wheel in such a proportion that they showed a colourless grey, then it would be possible to set the black and white swatches in the centre in such a proportion that they showed the same shade of grey. The areas of all the swatches (read off a scale on the circumference) would then constitute a colour-equation, for instance:

(a) $.37V+.27U+.36EG=.28W + .72\ Bk.$

Maxwell arbitrarily chose Vermilion (V), Ultramarine (U), and Emerald Green (EG) as his basic colours, emphasising that they differed in hue from those commonly thought to be primary colours, indeed that their choice had been in large measure arbitrary and that they were nothing more than "kinds of paint."[43] The value for white obtained when these colours were mixed he then normalised by a factor of 3.57=1/0.28 (where 0.28 is the quantity of white (W) determined in the above measurement), using this value to adjust the coefficients of all other equations he arrived at. When multiplied with the raw weights of (a), this factor yields the quantities of V, U and EG which would be required to yield one brightness-unit of W. So, for example, in order to locate Pale Chrome (PC) on the plane, he took the following equation, in which PC has been substituted for the basic colour V, and the proportions of U, EG and W have been adjusted to yield a match between two greys—the one resulting from the coloured swatches, the other from the black and white ones:

(b) .33PC+.55U+.12EG=.37W+.63Bk.

Multiplying through by the normalising factor 3.57, one gets:

(c) xPC+.55U+.12EG=1.32W

Maxwell then solved for x=1.32-(.55+.12)=.65, calling this result the "corrected value of .32PC."[44] It specified uniquely the location of Pale Chrome on the plane.

The need for this adjustment is obvious from the equations (a) and (b). The swatches V, U, and EG mix to form a darker white than do PC, U, and EG, as is reflected in the lower value for W (.28 as opposed to .37). Thus the effective brightness of the swatch of PC used must be greater than that of the swatch of V used in the basic colour equation (a). Normalising its brightness to that of V means increasing its coefficient accordingly. This adjustment was a direct application of Graßmann's fourth principle (the brightness of the mixed colour must equal the sum of the brightnesses of the component colours), which serves to emphasise just how much Maxwell's approach assumed the validity of that mathematical analysis. In essence, Maxwell chose the second of Helmholtz's interpretations of Graßmann's fourth principle: selecting three arbitrary colours, whose brightness units were implicitly determined in the choice of the kinds of paint representing them, he constructed the colour

43 "The reason for this deviation from the received system is, that the colours on the discs do not represent primary colours at all, but are simply specimens of different kinds of paint, and the choice of these was determined solely by the power of forming the requisite variety of combinations." (Maxwell 1855), p. 276.

44 (Maxwell 1855), p. 280. Maxwell means .33PC.

plane by defining the locations and brightness units of all other colours as functions of those six posited values.

One can visualise Maxwell's procedure as follows. Imagine that the colours form a roughly conic solid, whose axis runs from a pure white apex to a pure black point at the centre of its base. Cuts perpendicular to the axis represent colour planes of constant saturation radiating from a grey point to a circumference along which the spectral colours lie. Maxwell began by selecting three arbitrary points in this solid, which together define a plane cutting through it. He then determined the positions of these three points on that plane relative to the axis of the solid. Using the colour equations resulting from measurements with the top, he then situated other colours to construct a colour-plane which, like Graßmann's, was a projection of the surface of the cone onto that plane.

The mere fact that three colours sufficed to determine the shape of the plane suggested that three fundamental colours would indeed be sufficient to generate all the others in a manner consistent with Graßmann's analysis, but for one niggling point: no pure spectral colours could be synthesised—the purity of the mixed colours was always inferior to that of the spectral colours. And even if one ignored that fact, it remained the case that, as Helmholtz argued, the shape of the colour-plane (whose contour was the spectral colour curve) depended on the choice of the three basis colours used in a particular experimental set-up, along with their postulated units of brightness. So by carrying out measurements of this sort one would not, without further refinements, be able to figure out *which* three colours were most like, i. e., closest to, the true primaries. Maxwell disposed of these difficulties by taking two radical steps: first, he argued that the three true primaries, whatever they were, *did not lie on the colour-plane*; second, he compared the measurements he had obtained for the bulk of his subjects (where there was very good agreement) with those of the colour-blind subjects, and used these results to locate the elementary red-sensation.

What previous advocates of Young's theory had failed to understand, Maxwell argued, was that the three fundamental retinal processes, which Young had claimed corresponded to the three sensations of red, green and violet, were not identical either with those three colour-sensations, nor were they directly connected to "the lengths of the undulations of the various rays of light."[45] Each of these three "systems of nerves … acts not, as

45 (Maxwell 1855), p. 284.

some have thought, by conveying to the mind the knowledge of the undulation of light, or of its periodic time, but simply by being *more* or *less* capable only of increase or diminution…."[46] Every light-ray whose wavelength falls within the visible range acts on all three systems simultaneously, even if it is a pure spectral ray. The range of colours which we normally perceive can be represented as a figure inscribed within the triangle of possible "mixtures" of these three elementary colour-sensations (Fig. 4, the circle is Newton's wheel; the vertices of the triangle are Maxwell's elementary, or "pure" sensations).

Nonetheless, the exact location of the new "pure sensations"[47] could not be described by means of observations of colours lying within the circle, and that for two reasons, the one mathematical, the other physical. (1) Any number of non-collinear triplets of points lying outside the circle could form a basis for it (that is, any three that enclosed it), so any three such points might represent the three elementary sensations. In the terms of Graßmann's vector analysis, one could not draw conclusions about the vectors—thus the system of coordinates—which formed the physical basis of the colour-plane by considering its structural properties alone.[48] (2) If we had access to the physical correlates of the colours lying outside the wheel, but inside the triangle, we might overcome this difficulty. But the colours within the wheel were all and only those held to have a physical cause, that is to say, whose ultimate correlate was posited outside the mind and body of the subject. And Maxwell and Helmholtz could alter only *such* causes directly. The difficulties encountered in manipulating and measuring the elements of the colour-plane had been up until this point largely technical and definitional. But on the new theory, these elements had dropped out of sight for the subject, and had slipped from the experimenter's grasp.[49]

---

46 (Maxwell 1855), p. 283.

47 (Maxwell 1855), p. 295.

48 Each set of three vectors corresponded to an affine transformation of the colour-space, and the significant topological and metrical relations of that space were invariant under those transformations.

49 I am deliberately omitting the second innovation in Maxwell's approach: his use of colour-blind subjects. This did make it possible to draw conclusions about the elementary red-sensation—precisely because such individuals have, on the Young-Maxwell-Helmholtz theory, a two-dimensional colour-field. The introduction of other, pathological, subjects into the research programme changed it profoundly. To do that change justice would require a longer treatment, and one with a different focus.

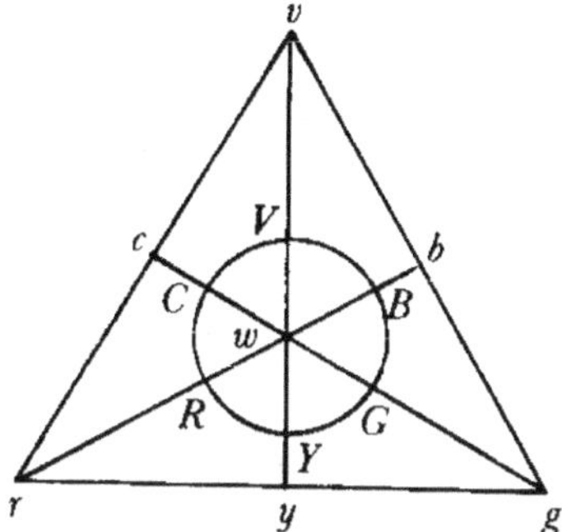

Fig. 4: Maxwell's extended colour-plane.

By the time the first edition of the *Handbook of Physiological Optics* was published in 1860, Helmholtz had also converted, without giving much credit to Maxwell, to exactly this view. His interest in the exact nature of the elementary sensations, and in the mathematical description of the colour-space to which they gave rise, continued until his death: the second edition of the *Handbook* contains an extension of the trivariance theory, in which the elementary sensations are explicitly identified as mathematical functions realised in the conductive nerve tissue;[50] and Helmholtz attempted, in a series of paper in the 1890's, to determine the "metric" of the colour-space.[51] What is more important to the topic of this study, however, is the fact that he repeatedly referred to the role of colorimetry in his later papers on geometry; indeed, as I shall argue in the following chapters, his final philosophical rebuttal of his Kantian opponent Land effectively recapitulates the arguments he directed against Graßmann when analysing the latter's fourth axiom. Since, however, I must reserve detailed discussion of those arguments to Chapter 6, in which we shall consider each of the geometry papers in succession, I shall content myself in the concluding section of this chapter with analysing the relation of this work to the simultaneous difficulties concerning the status of geometry that had emerged in the dispute with Clausius which we considered in the last chapter.

50 (Helmholtz 1911), pp. 341–343. The third edition cited here is a reprint of the first edition.

51 (Helmholtz 1891a, 1891c, 1891b)

(f) Conclusion

As I suggested in the Introduction, the existence of the colour-space was considered by virtually all writers on the space-problem, from Riemann onwards, to be of great significance. This fact may strike us today as peculiar, because we are accustomed to describing phenomena as forming a "space," whether we do so merely metaphorically, or when actually employing a formal description in which the phenomena are organised as a matrix or a continuum. The situation in the period we are considering was quite different, however. We have already seen in some detail that geometry, for both Kant and Helmholtz (and the same would hold for their contemporaries), is not a purely formal science. Even when we express them algebraically, the propositions of geometry refer to a single structure of our experience, namely physical space. Of course, it is once we have said this that the debate concerning the ontological status of physical space, and of the set of properties attached to it, first gets started. Kant follows Leibniz in reproaching Newton and his followers for reifying space. But he does not accept Leibniz's claim that we can do without spatial relations in the basic vocabulary of science, for he contends that space grounds relational properties that are ineliminable. Thus Kant attempts to steer a middle course, according to which space is neither a thing in itself nor a pure chimera. As a form of experience, it is objectively real in the sense of being a condition for all (human) experience of the world. But it nevertheless remains ideal, because it has no autonomous existence. The strict relativity of space is essentially a corollary of this doctrine. For if space is merely a form, it can only be experienced when it has a content. Thus there can be no relations between objects and space itself.

Now, as we saw in the Chapter 1, Kant's theory of spatial magnitudes runs into difficulties when he has to distinguish between empirically given spatial magnitudes, and those spatial magnitudes which, although they are not empirically determinate, nevertheless count as objects of geometry. I suggested in my concluding discussion that, from Kant's point of view, the problem does not arise because he thinks that the topological properties of the spatial manifold, namely its being a three-dimensional extensive magnitude, *entail* its metrical properties. It is the relations of part-whole containment between solids, planes, lines and points that characterise the extensive structure of space, and the axioms of geometry codify relations of equality between such regions of space. Thus while it is true that certain ideal mechanical operations are involved in geometric constructions, these operations do nothing more than verify relations of equality that already exist among these regions. When they are filled in with a content, the latter relations of equality continue to hold, and this by hypothesis: they are relations holding among parts of the container, and thus only secondarily between what the latter contains.

Matters are quite different when it comes to the other species of magnitudes, namely Kant's "intensive magnitudes." They are magnitudes in an entirely different sense: they can, according to Kant, increase and decrease in time, giving them continuity and order; however, they do not have a part-whole containment structure. Furthermore, we may assume (Kant gives no examples that suggest otherwise) that such magnitudes are exclusively scalar, or one-dimensional. The two kinds of magnitudes are thus distinguished (1) by the presence and absence of containment relations, (2) by their dimensionality. Furthermore, as Kant explains in the Anticipations of Perception, we must relate intensive magnitudes to the extensive magnitude of (at least) time in order to assign them quantitative measures. Space and time thus retain their status as the sole ground of properly quantitative relations. There is no sense in which mathematics could be dissociated from them, in order to become purely analytic and formal. Nor are there any other candidates to take on this semantic role.

The importance of the colour-space for this theory of magnitudes is somewhat easier to appreciate if one reflects on this last point. Helmholtz, like his teachers Lotze and Herbart, was originally committed to the theory I just described. The introductory sections of the *Conservation* make it perfectly clear that the task of natural science is to analyse all phenomena in terms of laws holding between spatio-temporal magnitudes, i. e., in terms of the paths of material points in the spatio-temporal manifold. The sciences of intensive magnitudes such as colours or auditory tones are no different. These should also seek to analyse such intensive magnitudes in terms of their mechanical causes. This project, conceived realistically, is simply reductive: since the causes of perceptions are assumed to be mechanical processes, it must be possible to relate all changes in appearances to these processes, where the latter can be mathematically described. Conceived transcendentally, the reduction is mandated by the regulative requirement that we represent processes mechanically *in order that* they may be described mathematically—there is no guarantee that such a reduction is possible.

Whatever one's preferred philosophical interpretation, the method is the same: Newton, Helmholtz and Maxwell all tried to quantify the relevant properties of spectral light on the one hand, and the observable relations of equality between colour-perceptions on the other. On the assumption that there *were* such relations of equality among distinct colours (for instance that two shades of colour can be equally bright), the task could be easily completed: one would lay down the colour plane by determining the dependence of that perceived brightness on light quantity, and of spectral colour on wavelength, respectively. But the research programme I just described gave rise to an unexpected result: there was no unequivocal sense in which the brightness of distinct shades could be compared with one another. Despite the properties of continuity and dimensionality that are inherent to the colour-space, the latter has no "intuitive" metric, although it does consist of intensive magnitudes that exhibit additive properties. Thus the colour-space presents a counter-example to Kant's theory. We are already acquainted with a continuum of *intensive* magnitudes which, although they do not contain each other, can be "added" by mixing them. And the "geometry" of this continuum is *not* determined merely because such an additivity operation exists. The reason that so many of the authors I mentioned in the Introduction (e.g. Carnap, Poincaré, Russell, Weyl, plus of course those discussed in this chapter) are fascinated by colours is quite simply this: they provide an alternative semantic for the theory of space, and this model makes it clear that continuity and additivity alone do not entail that a particular geometry is correct. Moreover, it is possible to say of one and the same continuum that it can have different metrical relations. In saying this, I do not of course mean to suggest that all these possibilities emerged in the course of the five-year research programme that I have just discussed. But Helmholtz's remarks in his later papers on geometry, and indeed the arguments of several of these papers, make it evident that this early research played an exemplary role for him.

Helmholtz only realised the difficulties concealed in the concept of a colour-space once Graßmann had presented his vector analysis of the additive relations underlying its structure. He took Graßmann to have erred in his interpretation of his own mathematics, however. For, according to Helmholtz's scholium of Graßmann's fourth axiom (that the brightness of a mixture was equal to the sum of the brightnesses of its components), there was an element of conventional choice involved in determining a colour-metric. One could either specify in advance the shape of the spectral colour-curve, and then choose definitions of brightness and a measurement procedure that would be compatible with this curve. Or one could specify the definitions and the procedure, in order to derive the shape of the curve from the experimental results. Helmholtz, as we saw, opted for the latter approach, and claimed that the curve he had generated was preferable. As we shall see in the next chapter, he retrenched from this position in later writings, for instance in the second edition of the *Handbook*, where he admitted that all linear transformations of the curve represented the same empirical information.

In order to avoid any misunderstanding, I should emphasise that Helmholtz's dispute with Graßmann did not in any sense involve a choice between Euclidean and non-Euclidean metrics, even if, as I mentioned previously, he did eventually do research on the concept of a "colour-metric."[52] I am not claiming that this work could or would have been sufficient to raise the question concerning the "correct" geometry of physical space on its own. I will argue in the following chapter, however, that this research impinged on the problems that had emerged in Helmholtz's research in physics, some of which we have already explored in the previous chapter. For even before the publication of the *Conservation*, Helmholtz had conceived of geometrical propositions as statements concerning the possible displacements of ideally rigid bodies. Thus he was already grappling with the same tension in Kant's theory of mathematics that I described at the conclusion of Chapter 1, a tension resulting from three postulates: (1) space is undifferentiated and cannot be an object of perception, (2) physical theory should involve only magnitudes that are empirically determinate, (3) geometrical demonstrations involve motions of geometrical forms. And Helmholtz had, in the dispute with Clausius, insisted on (2) for the purpose of disqualifying forces which involved relations to absolute space. His argument, as we may recall, assumed that any statements regarding the internal state of a system had to be interpreted as statements concerning relations among real things, from which it followed that one could not appeal to merely mathematical coordinate systems when defining physically applicable concepts.

This last debate occurred in the middle of the research programme on the colour-space that I have outlined in this chapter. Here as well, Helmholtz was confronted with the problem of defining the "proper" coordinate system for a continuous, multi-dimensional manifold. And here again, Helmholtz rejected the idea that one could determine the coordinate system of the space by stipulation, in order to adjust the procedures of measurement so as to conform to it. Of course, there is a significant difference between the notion of a colour-metric and that of a properly spatial metric, which consists in the simple fact that there are no events

52 *Cf.* foonote 51.

or processes "in" the colour-space. The matter of determining a correct metric is thus exclusively a question of establishing correlations between colours and colour-intensities that we perceive as equal, and the quantitative measurements of their physical causes. Both Helmholtz and Maxwell expected that these correlations would be explained by the actions of hidden receptor organs, as Helmholtz had managed to explain the properties of the "space" of auditory sensations in terms of the acoustical properties of the cochlea. But such an expectation does not touch the problems raised in establishing the correlation between subjective and objective properties.

In the case of space proper, these problems are still more acute. There are no external factors to correlate with those spatial relations that we intuit as equal. Helmholtz did, in his last paper on geometry, frame the question of the objective validity of geometry in just these terms, thus betraying the importance of the colour-space for his interpretation of his geometric arguments. But since he remains even at that point committed to the Kantian thesis that such relations are in principle beyond our possible experience, such an experiment remained a *thought*-experiment. Since we cannot compare space as we experience it with its noumenal grounds (which Helmholtz there calls "topogeneous moments"), the only way we can give these equalities an objective meaning is by correlating them with physical processes that occur within the spatial manifold itself.[53]

This is the lesson that, on my reading, Helmholtz learned in the research I have described in this chapter. While it is possible that we intuit equal brightnesses, or equal distances within a multi-dimensional continuum, these intuited relations do not have any objective significance until the relations they define—for instance the equivalence class of line segments of the same length—have been operationally grounded by means of a measurement procedure. Furthermore, the classes in question must be related to some further set of experiences, whether these consist of their mechanical causes (in colour-theory), or the set of inertial motions (in physics). Our intuitive estimation of such relations is *not empirically meaningful* until these last two elements are specified. In this sense, geometry does not yet make any objective statements about the world. Of

53 This of course will lead to a circularity, for it will mean checking spatial properties against themselves. Helmholtz addresses just this difficulty in the first paper in geometry (Helmholtz 1868a), and hopes to circumvent it by using purely analytical methods.

course, Kant would have agreed with Helmholtz to the extent that he maintained that geometry was not an empirical science. Helmholtz's take of the matter is somewhat different, however, for he will come to argue that the role of geometry can be entirely taken over by Kant's phoronomy. According to Helmholtz's final word on the matter,[54] we ought to replace pure intuitive geometry with what he called "physical geometry." Or, to put it in the language he used twenty-four years earlier in his reply to Clausius, it is only when intuitions have been connected to relations among real things that we are dealing with magnitudes that are not "merely drawn on paper."

54 In (Helmholtz 1878c).

5. The Road to Empirical Geometry

In the last three chapters, we have examined the sources of Helmholtz's views on the philosophy of science, the nature of basic physical concepts, and the theory of manifolds. Our concern in the concluding chapters will be to analyse his application of the theoretical insights that emerged in this work in his four seminal papers on geometry.[1] I will address the arguments of those four papers in detail in Chapter 6; however, before we turn to that analysis, we will take stock in this chapter of what we have learned so far. The general line of my argument will be that Helmholtz's work on geometry, although it was on his own admission motivated by his opposition to "nativist" brands of Kant's philosophy, cannot be properly understood in isolation from his work in physics and in measurement theory. Nor should it be seen as a brute rejection of Kant's philosophy of science in preference to a simple empiricism. There is no doubt that Helmholtz was active on all these fronts; however, physics and the philosophy of science were always his preferred domains, and it would indeed be surprising if he had overlooked the relevance of the philosophy of geometry to these domains. I shall not so much try to defend this negative thesis in the following discussion, however, as to urge a positive one, which I believe helps restore continuity between this work of Helmholtz's and its later discussion and application at the hands of Einstein and Poincaré. The positive thesis is that the theory of geometry was important to Helmholtz because of emerging tensions within physics, and specifically within electrodynamics.

This tension was the result of the first theories involving motion-dependent forces; for instance, the theories of Neumann and Weber mentioned in Chapter 3, as well as, in subsequent years, Maxwell's field-theory. Such theories forced a problem concealed within Kant's philosophy of science into the open. I addressed this concealed difficulty at the conclusion of Chapter 2. Kant had argued against Newton's postulate of absolute space, and defended instead, as indeed the general aims of the critical philosophy demanded, that space was merely a form of our intuitions. Since absolute space could never be the object of a possible empirical in-

1 (Helmholtz 1868a, 1868b, 1870, 1878c)

tuition, it could never be admitted as a concept of natural science.[2] But Kant remained committed to the constitutive role of geometrical principles in the metaphysics of nature. All natural scientific concepts had to be provided with pure empirical schemata that defined the range of intuitions that they subsume. Whereas intensive magnitudes such as force and velocity lacked the additive properties required for their employment in mathematical laws, these properties could be secured by "constructing" the concepts in the extensive magnitudes of space and time. Each higher-level concept would be provided with a schema which would "exposit" it on lower-level ones. At the base level, that of phoronomy, motive concepts were to be constructed geometrically, and the requisite additive relations would finally be grounded in the geometric relations inherent to the manifold of space itself.

The problem was that Kant remained committed to a Euclidean and constructive conception of geometry. Geometric proofs are essentially dependent on our ability to "draw" the mathematical objects that correspond to geometric concepts in pure intuition. Although these mathematical objects could not be thought of as empirical, they are in a sense the limit cases of actual empirical intuitions. They correspond to pure operations of the productive imagination, namely those operations involved in the production of any truly empirical representation, and thus the forms that they produce are characteristic forms of empirical objects. But this conception of geometrical forms is at odds with the project of the *Metaphysical Foundations*, for the latter demands of *all* of its concepts that they have an empirical content. There is therefore an unbridgeable gap between the theory of geometry and that of pure natural science. The problem only becomes vicious, however, once physical science begins to employ concepts that truly make reference to the attributes of absolute space. For, up until that point, both Kantians and strict Newtonians can offer divergent metaphysical interpretations of the same problematic thesis: geometry describes formal properties of something unobservable; however, we can suppress this apparent absurdity by arguing that, in any given situation, we can imagine geometry to be a description of some arbitrarily chosen "empirical" space. So long as we can suppose a local rest frame in which the Newtonian laws of motion hold, as both

2 Except, of course, in the sense explained in the Phenomenology of the *Metaphysical Foundations.* There Kant argues that the ideal concept of a rest-frame defined with respect to the centre of mass of the entire universe can substitute for our concept of absolute space.

Kant and Newton were willing to do, we can take geometrical propositions to be statements concerning the properties of that inertial frame. Geometrical concepts such as "straight line" and "distance" (the two which we shall discuss at greatest length in the following) are *relative* properties, and the rest frame can be defined relative to the masses with which we are immediately concerned.

Such a conception is no longer possible for a Kantian once relations to absolute space are admitted as legitimate. For then the entire argument deriving from the empirical determination of concepts draws a blank: if the requisite magnitudinal relations are truly to be conceived as relations between the parts of absolute space, then there can be no adequate schematisation of empirical concepts. Physics can no longer be an empirically closed system of the sort demanded by Kant. My suggestion in Chapter 3 was that Helmholtz was thrust on the horns of this dilemma once he attempted to parry Clausius's claims that non-central forces were conceivable. The virtue of assuming such forces was that just explained: so long as forces were central, there was no need to appeal to the properties of space itself in order to determine physical magnitudes, that is, render them observable. The mass-points in the system determined all the relevant magnitudes. Nevertheless, as I will explain in greater detail in this chapter, this argument rested on a conception of geometry that Helmholtz had developed on his own in the period before the *Conservation* was published. This theory apparently circumvented the problems in Kant's system because it employed analytic geometry to define basic physical concepts. Helmholtz, as we shall see, moved straight from algebraic characterisations of the spatial manifold to the kinematic and dynamic concepts employed in physics. The latter were intended to be empirically determinate, and Helmholtz advanced a series of arguments to secure this determinacy—arguments which are, however, from our point of view, redundant. In fact, the definition of congruence relations that Helmholtz employed was just as much dependent on an absolute coordinate system (or a family thereof) as were the concepts of non-central forces that he later denied to Clausius.

The upshot was the one I described at the end of Chapter 3: Helmholtz needed to explain how geometry could be interpreted as an "empirical" science, where by "empirical" we address not so much its inductive basis, as the fact that, as in Kant, the objects that it describes are to be interpreted as material objects. The model he eventually offered in support of this view was derived from the work on the colour-space, for it was only here that he became acquainted with a theory of magnitudes powerful enough to permit a distinction between the metrical and the topological characteristics of a manifold of intuition. And it was here that he encountered the possibility of linking the metrical axioms to operational definitions.

Thus, according to my presentation so far, Helmholtz's work on geometry prior to the first paper of 1868 can be divided into three phases, although this is admittedly a reflection of the sources that we have at our disposal. In this chapter, I shall take stock of each of these phases in succession, and I shall conclude with a summary of their relevance to the first "transcendental" argument for the empirical validity of Euclidean geometry offered by Helmholtz in his 1868 papers.[3] The first of these phases is that developed in the manuscript on "general physical concepts" reprinted by Königsberger, to which I referred briefly in Chapter 3. In this early work, Helmholtz insists already on the metrological content of geometry, in that he considers "mathematical objects" to be a species of ultimately rigid bodies that are used to determine relations of congruence between properly physical bodies. The framework remains, however, strongly Kantian: both the mathematical and the physical bodies are considered to be situated in a single space of pure intuition, and Helmholtz appeals explicitly to our ability to intuit the identity of mathematical bodies as they are displaced when explaining the presuppositions involved in the notion of rigidity. The second phase is the one that I described in Chapter 3, where Helmholtz first appeals to the distinction between "purely mathematical" and "empirically determinate" congruence relations in order to defend his arguments in the *Conservation of Energy.* As I suggested in the conclusion of that chapter, Helmholtz's position is at this point highly unstable. He follows Kant in arguing from the relativity of space to the form of possible force functions; however, the claim that "merely mathematical" spatial relations are physically inapplicable threatens to undermine his own definition of energy conservation. For he does not explain how exactly we are to distinguish between purely mathematical and empirically determinate spatial relations. However, there is no reason to suppose that he had more resources at his disposal at this time than those he had outlined in the early manuscript.

In calling this a second phase, I am therefore referring less to a specific doctrine that Helmholtz advances as I am to a specific problem. This is, in a word, the problem of explaining the connection between the determination of an inertial frame and the determination of spatial magni-

3 (Helmholtz 1868a, 1868b)

tudes that are to count as equal. Helmholtz, as we saw, is pushed into appealing to the empirical determination of coordinate systems in order to defend himself against Clausius's attack. By distinguishing between spatial relations determined by real things and purely mathematical spatial relationships, he is able to disqualify certain force functions. But, as I suggested, this puts the ball back in his court. For he is now implicitly committed to explaining which, if any, purely mathematical properties may be presupposed in our description of physical systems, and which are to be subjected to the further restriction of being empirically determined. For instance, he has excluded directionality from the class of determinate properties just as he has, presumably, included dimensionality among them. But his own formulation of the energy principle assumes that distance *is* a physically determinate property. The energy principle, which is supposed to represent a fundamental law of nature, only has a determinate meaning if it is meaningful to say of a system that it is in congruent states at different periods of time. So he still owes an explanation of what determines congruence.

The third phase is the one that we examined in detail in the last chapter, namely the extended research programme on the colour-space. This phase overlaps with the first two, in that the earliest papers on colour-theory are also from the 1850's, but we can safely say that the properly geometric—or manifold-theoretical—problems first emerge with Graßmann's 1853 publication, to which both Helmholtz and Maxwell respond in 1855. All in all, Helmholtz is concerned with the empirical significance of metrical relations on two fronts in the years 1854–1855, although these two areas have, on the surface, virtually nothing to do with one another. The most careful theoretical arguments are to be found in the colour-theoretical papers, particularly in the "scholium," as I called it, concerning Graßmann's fourth axiom. This fourth axiom stipulated that the intensity (the brightness) of a mixed colour be equal to the sum of the intensities of its component colours. Without this axiom, the combinatory relations of colours would not be representable on the plane at all. But even with this postulate, the "correct" arrangement of the colours is not yet given. Helmholtz distinguished between three possible interpretations of the axiom, which I shall call the "aprioristic," the "conventionalist" and the "empiricist" interpretations. The aprioristic one meant stipulating both the additive procedure and the spectral curve that were to hold, and accepting the brightness relations that these postulates induced. The conventionalist one involved choosing arbitrary basis colours and brightness values, and then determining the planar arrangement that followed from that conventional choice. Finally, the empiricist interpretation required an empirical procedure for determining what was to count as equal brightness for the spectral colours. One would then determine the plane that was consistent with the results of the measurements. This interpretation, in contrast to the conventionalist one, left open the possibility that the colours might fail to satisfy the requirements of additivity and transitivity implicit in the other axioms.[4] Helmholtz opted for this empiricist variant, while he criticised Graßmann for choosing the aprioristic one. The conventionalist method was then applied by Maxwell, and Helmholtz, in his later publications on the colour-space, came to endorse this approach as well.

As I shall explore in greater detail in the last section of this chapter, this scholium is significant because of the link it established between the axiomatic and operational components of the colour plane. More precisely, it made evident the fact that the determination of a particular metric depends on both axioms and the operations that realise them,[5] and the breach between the two was decisive for the interpretations given of non-Euclidean geometry for years to come. For although I am doubtless forcing the issue somewhat by naming the three interpretations as I do, the correspondence between Helmholtz's three interpretations of the colour axiom and later epistemological interpretations of geometry is no accident. Thinking about colours enabled Helmholtz and others to break the hold of Euclidean geometry on their imagination. In the case of Helmholtz, it enabled him to radicalise his own thinking about spatial measurement. For, as we have seen, Helmholtz had already been thinking about the role of rigid bodies in determining congruence relations before beginning work on the *Conservation*, in what I have called the first phase. And he had also applied the results of this thinking in order to parry Clausius's criticisms of his memoir. But it is equally clear that neither he nor Clausius thought that in demanding that coordinate systems be empirically constructable, and not merely mathematical, one might call the validity of Euclidean geometry into question. On the contrary,

4 Helmholtz's student Johannes von Kries later referred to this brightness postulate in order to illustrate the connection between additivity and transitivity postulates in the definition of magnitudes, (Kries 1882), pp. 268–270. *Cf.* (Darrigol 2003), p. 539.

5 Olivier Darrigol suggests that this fusion of "conventional and objective elements" is definitive of Helmholtz's theory of measurement in his much later work, *Zählen und Messen, erkenntnistheoretisch betrachtet* (Helmholtz 1887). *Cf.* (Darrigol 2003), p. 516.

they saw this as a claim concerning the distinction between relative and absolute spatial relations. But the two chains of reasoning, when taken together, lead almost inevitably to a question concerning the operations underlying the possibility of constructing coordinate systems which, as Helmholtz put it, depend only on the relations among real things.

(a) Phase 1

In discussing the *Conservation* in Chapter 3, I referred to an earlier manuscript concerning what Helmholtz called "general physical concepts."[6] The exact relation of this manuscript to the *Conservation* remains unclear, since Königsberger reproduces it only in part and does not give an exact date, saying only that it is from a few years before the publication of the *Conservation*. It is, however, significant to our understanding both of the *Conservation* and of the later papers on geometry for a number of reasons. As I shall explain in this section, even at this early stage, Helmholtz argues that the concept of congruence is not among those spatial concepts (or "determinations," as he puts it) which are necessary to space merely containing bodies. Rather it belongs to a second class of determinations, namely those necessary to its containing or comprehending (*umfassen*) *moving* bodies. Thus, to a limited degree, Helmholtz already subscribes to a thesis that he explicitly asserts twenty years later, namely that geometry treats of the motive properties of bodies.

Nonetheless, it would be misleading to suggest that his views are identical in these early and late phases. In this early manuscript, Helmholtz is concerned with giving *a priori* definitions of the core concepts of physics. The fact that some of these concepts are motive concepts in no way impugns their apodictic character. In fact, the philosophical explanation of what he calls "general physical concepts" (*die allgemeinen Naturbegriffe*) in what I shall call the Königsberger manuscript follows Kant's doctrines quite strictly. According to Helmholtz, the natural sciences are concerned with that portion of our representations which is not of our own making. Certain properties of these representations make up their "necessary form," in the sense that no representation at all can be formed without them. The system of concepts describing these properties will therefore form the *a priori* core of natural science:

6 (Königsberger 1903), pp. 126–138.

> These natural concepts are derived in part from the brute fact that there are determinate perceptions at all, which are not produced by our own activity, and in part from individual determinate empirical perceptions themselves. The system of the first of these yields the general or pure natural sciences (theory of time [*Zeitlehre*], geometry, pure mechanics), and the system of the second yields theoretical physics. The common feature of the general natural concepts will be that they and their consequences are the basis of all natural intuition [*Naturanschauung*], thus that they are in this regard the general and necessary form of natural intuition, thus also that the certainty of their propositions is an absolute one, whereas the certainty of the specific natural concepts only ever extends so far as to say that all facts known up until now agree to them. Furthermore, the general concepts, derived only from the possibility of any natural intuition, may not restrict the possibility of any empirical combination of perceptions, i. e., no empirical fact or law may be derivable from them, rather they can yield only a norm for our explanations.[7]

The task that Helmholtz sets himself in the rest of the manuscript is thus essentially the same as that undertaken by Kant in the *Metaphysical Foundations:* to define the *a priori* principles of the natural sciences by a "pure empirical" analysis of the concepts of time, space, matter, and force. Just as Kant argued that these pure empirical concepts were valid because they derived from constitutive categories and the pure intuitions of space and time, so Helmholtz considers them "certain" and "absolute" because they describe the conditions on physical (or "natural") representations. But his methods are from the outset distinguished from those of Kant by the use of algebraic techniques. As in the introductory sections to the *Conserva-*

7 (Königsberger 1903), p. 127. The German reads: "Diese Naturbegriffe werden erschlossen, teils aus dem Factum allein, dass es überhaupt bestimmte Wahrnehmungen gebe, die nicht durch unsere Selbsttätigkeit hervorgebracht sind, teils aus einzelnen bestimmten empirischen Wahrnehmungen selbst. Das System der ersteren gibt die allgemeinen oder reinen Naturwissenschaften (Zeitlehre, Geometrie, reine Mechanik), das der letzteren die theoretische Physik. Das Gemeinsame der allgemeinen Naturbegriffe wird sein: dass sie und ihre Folgerungen aller Naturanschauung zum Grunde liegen, und ohne sie keine gedacht werden kann, dass sie also in dieser Hinsicht die allgemeine und notwendige Form der Naturanschauung sind, daher auch die Gewissheit ihrer Sätze eine absolute ist, während sich die der besonderen Naturbegriffe immer nur so weit erstreckt, um auszusagen, dass alle bis jetzt bekannte Facta ihnen entsprechen. Die allgemeinen Begriffe, nur hergeleitet aus der Möglichkeit irgend einer Naturanschauung, dürfen ferner nicht die Möglichkeit irgend einer empirischen Combination von Wahrnehmungen beschränken, d. h. es darf aus ihnen durchaus kein empirisches Factum oder Gesetz ableitbar sein, sondern sie können uns nur eine Norm für unsere Erklärungen geben."

tion itself, many typically Kantian doctrines recur in analytic garb. The immediate consequence of this new technique is that Kant's notion of a geometric "construction" has no place in Helmholtz's manuscript. In the following, I shall confine myself to that portion of his discussion which is concerned with the distinction between geometrical and mechanical principles, or in Helmholtz's reformulation, between non-motive and motive spatial relations.

After treating the properties of time, matter and cause, Helmholtz turns to the concept of space, which he characterises as a three-dimensional unbounded continuous magnitude. Each indivisible element of this magnitude can be determined by means of three determination-bits (*Bestimmungsstücke*), that is to say by three coordinate-values. Complex bodies and systems may then characterised by the "equations of determination" that specify the dependencies between the determinations of their parts: a line requires two such equations of determination, a plane requires one. Helmholtz then considers the fundamental determination of one point in relation to another, namely the notion of a distance, defining the latter as the length of the shortest line between two points. These various definitions are all those required to specify how a representation may be enclosed or comprehended by space (*umfasst*), and so Helmholtz turns in the second half of this exposition to determining the concept of space "in such a way that it can comprehend all possible changes of matter, which are obviously to be considered here only as changes in spatial relations, i. e., as motions."[8]

Most striking about this presentation is that Helmholtz defines *congruent* and *straight line* only when he turns to motive spatial relations. He has already, in other words, shifted concepts which, on the classical view, are properly geometric, to the kinematic or phoronomic domain. The most basic determination of two points—their distance—is defined metrically, and it is still a property of space that is required for it to comprehend motionless bodies. But while congruence is indeed explained in terms of invariant distance, this invariance is, as Helmholtz emphasises, the invariance of the relative positions (the distances) among the elements of a body *as the latter is moved* to coincide with the body with which it is congruent. Similarly, a straight line is distinguished from all other lines with reference to its determinative relations to other points: two points on the line determine all the others, and no points off the line are uniquely determinable by means of a point lying on it. Nevertheless, the concept

8 (Königsberger 1903), p. 134.

of a straight line is one that is first required in order to define inertial motions, and thus forces, and it is not among the properties of space that Helmholtz requires for the comprehension of motionless bodies.

Although Helmholtz's analytic approach frees him to define these motive concepts by means of "equations of determination," this method leads him to overlook just those assumptions which he later problematises in his work on geometry. For example, he defines length by means of the metric implicit in the coordinate system. From this definition, he can derive his concept of distance as a "shortest line," in order to use this notion to define rigidity as invariant distance through displacement. He again makes concealed use of the coordinate-system when he defines "straightness" by appealing to what we might term the *symmetrical indetermination* of straight lines: a straight line is, so to speak, indifferent to the sets of points on either side of it. But this concept is ultimately explained by referring to the properties of the "equations of determination" of a line, and thus, once again, to the properties induced by the choice of the coordinate-system. Helmholtz's definitions of these motive concepts hinge on the concepts of congruence and of straightness. His reasoning and terminology are very obscure, and I can only offer a charitable reconstruction:

(1) A rigid body is one in which the relative positions of its component points do not change as the body is displaced. By means of this definition, Helmholtz then defines the notion of congruence: two rigid systems are congruent if they can always be made to coincide.

(2) The path of a system is defined as the set of positions that the system runs through when it undergoes a continuous displacement. This path can therefore be determined by those equations whose values correspond to the points in question. Finally,

(3) The concept of an inertial, or straight-line path is broken down into two subdefinitions:

(3a) The direction of a path is defined as the "determination of a motion through which those points are laid down [*festgesetzt*] through which the motion would pass if continued without disturbance" (i. e. without the action of a force).

(3b) The "form" of an inertial path is defined as one whose direction does not change.

This last subdefinition is then given a specific geometrical correlate by means of a rather idiosyncratic definition of a straight line. Helmholtz requires that each section of an inertial path must be congruent to each other equally long section, for otherwise it would make no sense to think of the motion as uniform. Since this requirement is consistent with a path of constant curvature, it does not determine the requisite concept fully. But, Helmholtz argues, according to (3a) the form of the path must be entirely derivable from the "state of motion" of the body, which he has postulated as being "stable" or invariant. This "state of motion" is the cause of the progression of the body (*der Grund des Weitergehens*), and it must be in a given direction. Thus the path we are looking for is one whose direction does not change. Accordingly, such an inertial path must satisfy the following condition: every point that does not lie along the path must be symmetrically indeterminate. This means both that: (a) no subset of points on the path may permit the complete determination of a point that is not on the path, and (b) if a point not on the path is determined with respect to any two points on the path, then it is determined with respect to all the others. From this definition, Helmholtz concludes (without further argument) that there is only one straight line joining any two points, that equally long straight lines are congruent, and that "as a result, a straight line can be measured by transporting one of its parts which has been designated as a measuring unit."[9]

As in the arguments for force centrality in the opening sections of the *Conservation*, Helmholtz attempts in this manuscript to extract an extraordinary number of propositions by repeated appeals to what I have earlier called the principle of positional determinacy, as well as to the principle of sufficient reason. Inertial motion is unchanging motion, thus it is motion along a line that is characterised by uniformity. No parts of the line are specially determined, meaning that its equally long sections are congruent, and it does not determine points to either side of it. As in the case of his definition of congruence, however, this argument from *in*determinacy makes a concealed appeal to relations defined in terms of the coordinate system. In both cases, Helmholtz is attempting to give definitions of the pure empirical concepts involved in physical theories *without* assuming the classical geometric objects and constructions. But he fails to see that his analytic descriptions determine essentially the same notions of distance and straightness that he is trying to define by means of his appeals to "determinacy." Indeed, from our point of view, the entire metaphysical apparatus employed here seems largely superfluous.

Thus in stark contrast to Kant's theory of space, in which geometrical concepts such as those of a straight line or of congruent magnitudes are sharply distinguished from the "pure empirical" concepts of the *Meta-*

9 (Königsberger 1903), p. 137.

physical Foundations, Helmholtz's treatment of space apparently distinguishes only between determinations that are required for motionless and moving bodies, respectively. This distinction does not coincide with the traditional division between geometry and pure mechanics, since congruence relations are assigned to the motive side. Conversely, the class of spatial properties that are forms of motionless appearances is more restricted than it is in traditional geometry. Helmholtz can only proceed in this manner because his definitions of these general concepts of nature are from the outset analytical. Here, as in the papers of 1868, Helmholtz sees the truths of analytic geometry as being analytic in Kant's sense: the properties of space that one describes by means of equations are true by virtue of their conceptual content; as a result, they do not describe proper determinations of the space of intuition.[10] They are, in other words, not in themselves adequate to the needs of natural science, even if they may be used as an aid to specifying the general concepts of nature. In consequence, Helmholtz supplements his analytic descriptions by means of the notion of positional determination: the featureless space of intuition must be parcelled out in specific magnitudes in order for spatial concepts to have a determinate application. And yet, despite the power of this method, indeed in part because of it, Helmholtz does not see that some of his basic analytical propositions have a geometrical content.

In sum, Helmholtz's account of spatial relations prior to the *Conservation* already assumes that geometrical constructions involving congruence fall within the province of what Kant called phoronomy. For they depend on the motive characteristics of rigid bodies. Even if the latter are *ideal*, they still represent a limit case of material bodies, and the spatial relations they define are those involved in kinematic statements. Despite this attempt to fuse geometry with phoronomy, Helmholtz does not truly attain his goal. His definitions of a straight line and of congruence are related to the notions of an inertial path and of empirical measurability by means of metaphysical expositions. For instance, an inertial path is

10 Obviously Kant would also have regarded the propositions of analytic geometry as synthetic as well, for they deal with numbers, and all mathematical truths are synthetic. But even in (Königsberger 1903), p. 128, Helmholtz describes the truths of arithmetic as depending on the laws of logic plus the basic concepts of quantity, quality, equality and comparison; furthermore, he introduces these notions before, and without any explicit reference to, the intuition of time. In other words, he seems not to have viewed algebraic truths as synthetic *a priori* in Kant's sense.

an *unchanging* one, and this property is reflected in the symmetrical indeterminacy of the points lying on either side of the line—there could be no sufficient reason why a particle would deviate in the one direction or the other. But here, just as in the definition of rigid bodies as ones whose internal connections are invariant, Helmholtz must appeal to relations that derive from the coordinate system. He views this description as in some sense neutral, overlooking the fact that the choice of this system effectively determines what will count as straight lines and distances. And thus the arguments resting on the "determinacy" of distances and directions are essentially superfluous at this stage.

(b) Phase 2

We have already discussed Helmholtz's application of some of these principles in a physical context in Chapter 3. In defending his philosophical arguments in the *Conservation of Energy* against Clausius's objections, Helmholtz extended their application beyond the purely epistemological domain, and used them to invalidate physical theories that employed what he called purely mathematical magnitudinal concepts. The essential point, we may recall, was that whereas purely analytical descriptions of space permitted us to define arbitrary directions, the basic magnitudes of physical theories should all refer to possible experiences. These possible experiences had to be empirically determinate if nature was to be completely comprehensible. Helmholtz's objections gain some legitimacy in the light of the manuscript we have just discussed, for it makes clear that Helmholtz was no opponent of analytical methods, nor did he invent the distinction between merely mathematical and real spatial relations merely for the purpose of rebutting Clausius. The aim of this early manuscript on general physical concepts could accurately be described as that of providing determinate definitions of basic physical concepts. That is indeed the function of the superfluous "determinacy" arguments, which do not in fact go beyond the content implicit in the analytic descriptions.

Nevertheless, as I argued at the conclusion of Chapter 3, Helmholtz's rebuttal of Clausius was inherently unstable: he had to deny that directional properties were definable with reference to mathematical coordinate-systems all while maintaining that congruence relations did not depend on such reference. Now, as we have just seen, the concept of congruence is defined in the Königsberger manuscript by means of the notion of a movable rigid body. A rigid body is one whose parts do not change their relative positions as it is transported. And the notion of a relative position is cashed out by means of the notion of a distance (*Entfernung*), which is the length (*Länge*) of the shortest line connecting two points. Two points, Helmholtz argues, determine only one kind of spatial magnitude, namely a line. Lines have *lengths*; however, there is no upper limit to the length of a line connecting two points. Helmholtz concludes that among all those that can be drawn between the two points, (at least) one line must have the shortest length. The length of the latter is the *distance* between the two. This latter property is a property of the point-pair, and so it can now meaningfully be used to define a rigid body. As we shall see in the next chapter, Helmholtz's first papers on geometry use essentially same approach: he defines rigidity analytically, with reference to algebraic relations on the coordinate system, in order to interpret congruence as a relation between rigid bodies.

At this point, it may seem to go without saying that the congruence of two systems is an empirically determinate property of the pair, whereas absolute direction is not. The reasoning would be a simple symmetry argument: if we consider only the empirically given points, then only those directions singled out by the lines connecting them are observable; that is to say, we cannot distinguish between "rotations" relative to absolute space. The problem, however, is not that this claim is false. The difficulty lies rather with the opposing claim that congruence relations are determinate in a stronger sense. For the definition of congruence by means of rigid body displacement also leads us back to a notion of length that is defined with reference to a given coordinate system, or at least a family thereof. Thus, although this weak point in his argument is evident neither in the *Conservation* nor in the reply to Clausius, Helmholtz has employed the same sort of "arbitrary" coordinate system that he denied to Clausius in order to define the notion of a distance. And this means only that the very metrical function that he later came to problematise in the papers of 1868 is tacitly assumed in order to define congruence.

In other words, if Helmholtz were to retain the distinction between purely analytic and mathematical properties of a space such as arbitrary directions, and those which have a claim to being "physically applicable" he would have to eliminate just this appeal to the primitive notions of *length* (a property of lines) and *distance* (a property of pairs of points that is defined as shortest length), which he used in the Königsberger manuscript to define rigid bodies and congruence. One can, however, invert the sequence of definitions, so as to define equal lengths in terms of congruence, and then to operationalise the definition of congruence. Two regions of space are then congruent if and only if they can contain one and the same rigid body. The concept of a rigid body is then, however, the primitive one. Rigid bodies are just those which preserve the same distance relations among their elements as they are displaced; however, there is no independent definition of what it means for distance relations to be preserved. To say that a body is rigid is to say nothing more than that it belongs to a class of bodies that can always be superimposed. This is essentially the approach that Helmholtz applied, with variations, in all four of the papers on geometry. I am not of course suggesting that he had made this step in 1854—the only point to retain here is that he had a pressing need to provide a definition of this sort that derived directly from the transcendental logic of the *Conservation*, and from his attempts to block the introduction of indeterminate magnitudes into the basic vocabulary of physics.

(c) Phase 3

There is no documentary evidence prior to the papers of 1868 to demonstrate when Helmholtz drew this conclusion—but he did draw it at least then. Nevertheless, there is ample evidence that he was already aware of the possibility of an operational definition of the concept of a magnitude as early as 1855. This evidence is to be found in the papers on the colour-space up to and including the first edition of the *Handbook of Physiological Optics*, in which Helmholtz was indeed forced to investigate the conditions under which a set of phenomena could be called a space. In the early papers, the question of whether colours were a space or not was basic: could one, on the basis of a few elementary colours, arrange the others as a continuum that was not merely intuitively continuous, but which also reflected the quantitative results of measurement operations? Graßmann's analysis, along with Maxwell's and Helmholtz's responses to him, sharpened the focus of the dispute considerably.

In setting out a mathematical method for describing colour mixture, and supplying at least preliminary physical and physiological interpretations of that method, these papers formed the basis on which all subsequent work on colorimetry drew. As such, they mark a watershed in the history of sense-physiological research. But all three men were equally well interested in the more general problems raised by this work, which centred on the correct methodology for determining the mathematical structure of a perceptual manifold. Indeed, I believe it is fair to say that, at least for Graßmann and Maxwell, this was the primary motivation. In the course of this investigation, it had emerged that the identity of this structure resulted from an interaction between definitions of "line" and "distance" on the one hand, and procedures and conventions implementing these definitions on the other. Such an interaction arises because phenomenal colours are not extensive magnitudes. They do not contain one another as proper parts, and thus there is no intuitive sense in which they can be added and subtracted. In consequence, the *phenomenal* colour-plane has no inherent metrical characteristics. Since this dissociation of metrical and topological properties is definitive for both Helmholtz's and Riemann's interpretations of the latter's theory of manifolds, it is unsurprising that the deliberations of our colour-theorists anticipate the methods and arguments employed in the later debates on the nature of another perceptual continuum, namely physical space itself. This point can be made evident by considering Helmholtz's analysis of Graßmann's fourth principle, and comparing the distinct possible interpretations he identifies there with his views—and the views he ascribed to his antagonists—in the later papers on geometry.

Considered formally, Graßmann's principle stated that the total intensity of a colour (the brightness of the spectral colour and the brightness of the admixed white) is equal to the sum of the total intensities of the colours out of which it is mixed. It is only on the strength of this principle that a colour-*plane* can encode all the relevant relations between the three colour-variables. It does so by making the brightness of white light a linear function of spectral brightness, and thereby permitting a projection of a three-dimensional surface onto the plane.[11]

Helmholtz's analysis of this principle showed, in essence, that sets of possible colour-planes were equivalent if they could be regarded as projections of the same surface. Graßmann had assumed that the surface in question was a cone, whose apex was white at a unit intensity, and whose base was a circle with the pure spectral colours radiating from a black centre to unit intensities at the circumference. The group of possible projections was therefore made up of elliptical curves.[12] Colour-planes arrived at by means of empirical measurements which did not belong to this family were incompatible with it, because they could not be projections of the same three-dimensional surface. Helmholtz thus objected

11 *Cf.* (Sherman 1981), p. 100.
12 (Helmholtz 1855), p. 69.

that Graßmann's *a priori* insistence on a plane with a given boundary could not be justified.

Of course, given a suitable adjustment in one's measurement procedure for the colours entering into a mixture, *any* planar arrangement consistent with the phenomenal continuity of the colours could be constructed. What Helmholtz objected to in Graßmann's paper was that the method he had proposed for determining quantities of colour could be assumed to agree with a particular disposition of colours on the plane, or groups of linear transformations thereof. There is some irony in this, since Graßmann—at this point the only person on the planet with a firm grasp of vector algebra—would certainly have got the point. What Helmholtz had shown was that planes which were linear transformations of each other could be regarded as empirically equivalent, whereas those which were not—those in which the affine structure of the space was not preserved—could not be so regarded. Since a line on the colour plane is defined by two colours lying on it, which in turn can be mixed to yield the other colours on the line, a transformation which destroyed these relationships would entail either a change in Graßmann's axioms, or in the specification of the measurement procedure, if it were to retain empirical validity.

In any event, Helmholtz was convinced that *he* had discovered something important: that the organisation of the continuum of colours was induced by the combination of three factors. The first of these was a definition of a straight line as a set of colours, any one of which can be mixed out of any two others in the set. The second was a definition of distance: a colour's location relative to two others on a line is given by the ratio of the quantities of the latter entering into the mixture. The third was an operational definition of unit quantities, that is, a method for measuring them. As he observed in the 1868 "Tatsächlichen Grundlagen der Geometrie,"[13] in which he compares the determination of distances in physical space by means of the displacement of rigid bodies to the definition of distance on the colour-plane, it was not a foregone conclusion that one would find a method that was consistent with these definitions. After all, he had failed to do just this in 1852. Furthermore, once one had settled on such a method, it was a matter for investigation to determine what arrangement the colours would take on that method.[14]

13 (Helmholtz 1868a)
14 (Helmholtz 1868b), p. 616.

There are two aspects of this analysis that are of particular significance. The first is the availability of an *appropriate* measurement method. The second is the *empirical equivalence* of groups of colour-planes. Before Graßmann's seminal paper, a host of planar and spatial arrangements of phenomenal colours had been proposed, many of which were intended to encode mixture relations (for instance of dyes, paints, threads in tapestries). Both the dimensionality and the continuity of the colour-"space" were well known. In his paper, Graßmann defined an axiomatic system that described how to tie such a spatial (in the reduced case, planar) arrangement to numerical values derived from a measurement procedure. But the definitions of brightness, colour-tone and saturation he gives do not presuppose a single physical interpretation. Much in the manner of modern measurement axioms, they define an ideal set of relations between colours, which may or may not be realisable in practice. Helmholtz's criticism of Graßmann would perhaps have been obvious to the latter, but it did identify an important point. Treating colours as *magnitudes* is contingent on being able to manipulate them in accordance with additive axioms, and the mere fact that they exhibit the topological characteristics of a spatial continuum is no guarantee that one will be able to do this in practice.

The second point concerns the groups of empirically equivalent planes resulting from modifying the placement and units of basis colours. In his discussion of Graßmann's principle, Helmholtz distinguished sharply between selections of basis-colours and units which *arbitrarily* determined the shape of the spectral curve (thus the arrangement of all colours on the plane), and the method he had used, which had made these determinations with reference to the equal brightness of pure spectral colours "as they appear to the eye." But Helmholtz weakened this claim in the successive editions of the *Handbook of Physiological Optics.* We have already seen why. The procedures used for constructing the plane cannot uniquely determine a single correct one, for the shape of the plane is always dependent on certain initial conventions—the choice of basis colours and their brightness units. These various planes can be imagined as projections of a conic surface on variously angled planes. Alternatively, as Helmholtz explained in the second edition of the *Handbook of Physiological Optics*, "if one imagines the geometric colour-plane as represented on an elastic plate, which one can stretch evenly in every direction, one would get a similar array of form-changes [*Gestaltsveränderungen*], but each of these colour-planes would remain correct."[15]

He goes on to ask, "What remains constant in these changes ?" Here there are two possible answers: (1) Even as brightness units are redefined conventionally, the affine structure of the plane remains the same; that is to say, given two distinct pairs of colours, the colour lying at the intersection of the two lines connecting each is always the same. Here there is no question of one or the other projection being empirically correct. (2) Alternatively, if brightness units and standard colours are simply *defined* in terms of objective properties of light, then a unique plane and arrangement of colours can be singled out. Helmholtz's (and later Maxwell's) curve of spectral colours was empirical in one respect: names of colours were replaced by wavelength values. Brightness was still subjective, however. A fully objective curve in the sense of modern chromaticity measurement replaces brightness by luminous flux. Here there is no more question of varying the curve, because the intensive magnitudes of colour phenomena have been completely replaced by extensive ones from another theory. If one sticks to the phenomena, however, the shape of the curve will necessarily depend on metrological conventions.

These alternatives represent two different ways of understanding what is "objective" in the structure of phenomenal experience. From the point of view of (1), there are relationships between colours that hold independently of our knowledge of the absolute ground of these relations. Some of these—continuity, dimensionality—are inherent to the phenomena. Others—their organisation in "lines" determined by mixtures—are the product of axioms stipulating that they be so arranged, and of the results of carrying out the measurements prescribed by those axioms. Finally, properties such as the distance separating the colours (what is to count as equal distance between colours) are determined by conventions for measuring quantities of the latter.

From the other point of view (2), all of these characteristics can be given a unique objective ground in terms of the physical characteristics of their underlying physiology, for instance in terms of exact response curves correlating excitations of elementary receptors to the wavelength and amplitude of physical light. This kind of objectivity removes any dependence of the colour-plane on conventional procedures, but only at the cost of replacing our terms for the phenomena with terms belonging to another physical theory (green becomes a wavelength of light, brightness

15 (Helmholtz 1896), p. 338.

a measure of radiation). The imperfectly realised quantitative relations between phenomenal colours take on an objective meaning either when they are tied to other phenomena by means of measurement operations, or when their physiological basis has been exhaustively (and mathematically) elucidated. Considered as a space of mere phenomena, the continuum of colours is nothing more than "a system of differences," to use Helmholtz's paraphrase of Riemann's definition of a manifold.[16]

How then did this research programme affect the theory of spatial measurement that Helmholtz had begun in the Königsberger manuscript? The critical factor was Graßmann's new vector calculus. Graßmann treated each possible colour-percept as an individual magnitude that could be arbitrarily added to the others (although not necessarily subtracted from them). The question of whether colours make up a space, and if so what sort of space, thereby became the question of determining the additive relations holding among these vectorial quantities. This approach was decisive for Helmholtz's understanding of physical space because it represented an alternative way of conceiving of distance on a manifold. As we saw, Helmholtz's analysis of spatial magnitudes differed from Kant's both in his analytic approach, as well as in his partitioning of geometric and motive concepts. Congruence was defined as a motive concept, but this definition made appeal to the notion of length, which was finally cashed out as a function on the coordinate system. But Helmholtz argued against Clausius that one could not make such an appeal to the coordinate system when defining physically meaningful, or determinate concepts. In doing so, he remained true to the properly Kantian part of his conception, which considers spatial magnitudes as forms of empirical appearances, and not as independently existing properties. Thus he was thrust on the horns of a dilemma: admit coordinate systems that ground relations between the points of a space independently of any given "determination"; or, stick to the Kantian picture according to which a spatial magnitude is only determinate when one is dealing with (at least) a pair of empirical points. Choosing the latter alternative would, however, mean bracketing both analytical and traditional axiomatic methods when analysing space physically. Let me explain this briefly.

On the modern approach, we define a metric space by specifying a topology, a coordinate chart (or system thereof) and a metric function. Once these definitions have been given, we can derive theorems concern-

16 (Helmholtz 1870), p. 16.

ing what Helmholtz and Kant would have called spatial *magnitudes:* in general, any collection of points in this space, but above all geometrical figures composed of lines, planes and solids. Although this apparatus was not available to Helmholtz, he was proceeding similarly in the Königsberger manuscript. He assumed a coordinate-chart, and defined spatial relations algebraically. But he still held that such definitions were not yet physically meaningful, because such an algebraic description is just that: a description of something that must already be given in intuition. The space of human intuition is not a freely created formal structure, but a form of possible experience. Indeed, since Helmholtz apparently held that the statements of analytic geometry were true *by definition*, no properties defined in this manner could count as "general concepts of nature." Thus the accusation levelled against Clausius was not an *ad hoc* complaint. It merely reflected the demand that any algebraic property of mathematical space must be given a correlate which characterises a possible experience. Whether or not Helmholtz would have put it that way, he was demanding that algebraic properties must be related to empirical schemata if they are to be physically significant.

This demand corresponds quite exactly to Kant's requirement that one "construct" the pure empirical concepts of the metaphysics of nature in intuition. But Kant, in contrast to Helmholtz, thinks that the same is required of geometrical concepts themselves. Kant makes this demand in part because he thinks that purely conceptual relations are inadequate to ground the truth of mathematics. The "construction" he requires has its natural correlate in Euclid's postulates, which "require" or "request" that we *be able* to construct figures of a particular sort. This is a style of geometry more congenial to the sort of philosophical argument that Helmholtz directs against Clausius: geometry is always concerned with particular figures "drawn" in space, and not with the properties of space itself, whatever they might be. From the demand that figures be constructed, it follows quite naturally that geometry is always concerned with determinate magnitudes (the parallel postulate being the notorious exception). But traditional axiomatic geometry is of limited value when we are doing physics. And this is not the result of its being less powerful than analytic geometry. The problem is that precisely because it does *not* describe the properties of space itself, it severely limits our abilities to define things like inertial paths and distances in empty space. This tension between the two geometrical interpretations was amply evident in the alternative interpretations of the parallelogram proof that I discussed in Chapter 3. From Newton onwards, the physical-analytical approach is favoured precisely because it frees us to characterise completely arbitrary spatial figures as subsets of real triples. The traditional notion of a magnitude—roughly, of something that is *in space* while being at the same time distinct of it—becomes superfluous.

The dilemma that Helmholtz faces is therefore the following: either give up on the notion of empirical determinacy as a constitutive requirement on geometrical statements, or give up on the power and generality of analytical tools. The first option gets us access to analytical methods, but it commits us to a transcendent conception of space, in which geometry describes properties of something that we can never experience directly. The second option secures the epistemological determinacy of geometrical concepts, but it forces us to regard a large part of geometry as depending on the existence of constructive procedures. And then the ontological status of the mathematical objects becomes typically problematic: Are they a species of physical object? If so, are they not also subject to the distorting action of forces? If not, how could they possibly be compared with physical objects in order to do physics? In light of these unattractive alternatives, the allure of Graßmann's vector mathematics is evident enough: it allows us to characterise space by means of the additive properties of vectors. Each vector can be immediately interpreted as representing a magnitude, while the space of analytic geometry remains open to us with its full complexity. The notion of a constructive procedure is retained: it corresponds to the additive operation on the vector space. But since the vectors themselves can be fully characterised in terms of analytical operations on the n-tuples characterising them, we can continue to use analytical methods to characterise their relations.

These analytical and philosophical virtues were already displayed in Graßmann's clean division between the three theoretical components entering into his characterisation of the colour-space: (1) the axioms defining the basic magnitudes and the laws of their combination, (2) the specification of the physical or intuitional data corresponding to these, and (3) the specification of the physical process corresponding to the combinatory axioms.[17] The question with which we began this section—wheth-

17 This division is preserved in modern measurement theory, in large measure because that theory derives from Helmholtz's original analysis in his *Zählen und Messen* (Helmholtz 1887). A comprehensive discussion of that work can be found in (Darrigol 2003). In the present context, thirty years before the publication of that work, Helmholtz was still learning the approach from Graßmann, albeit without acknowledgement.

er and in what sense the colours form a space—was thereby precisely stated for the first time: they form a space just in case the magnitudes specified in (2) satisfy the axioms (1) under the operations specified in (3). Helmholtz's criticisms of Graßmann were that Graßmann had assumed the existence of a measurement operation (3) that would satisfy the axioms (1) under a given of assignment of the basic data of (2). This is what I called above the "aprioristic" interpretation of the principle. Helmholtz argued for the "empiricist" interpretation, according to which a quite different colour-plane would result if the assignment specified in (2) was chosen "empirically." Lastly, Maxwell had followed the "conventionalist" approach, according to which both (1) and (3) were postulated, and a minimal set of basic colours consistent with these assumptions was arbitrarily laid down (2).

Now, according to my reconstruction of Helmholtz's views on space in 1854, every spatial magnitude, in order to be determinate, must be defined by at least two empirical points. Analytical methods are permissible in theoretical work, but all the magnitudes referred to in such analytical descriptions must be determinate in this sense if physics is to be applicable to nature. Well before his quarrel with Clausius, Helmholtz had made the possibility of measurement dependent on the concept of congruence, which he had defined as a motive concept. He had in effect superimposed two systems of spatial description: the one was purely analytical, although physically indeterminate; whereas the second, while indeed physically determinate, nevertheless contained a definition of shortest length that depended, covertly, on the physical applicability of the first. Even at this stage, however, Helmholtz's theory of space admits the possibility that one might live in a Euclidean manifold in which no objects existed by means of which congruence relations could be established. Or, to anticipate somewhat the arguments of his later papers, it is conceivable that the objects that existed might be freely transportable along straight lines, without it being the case that they could be rotated, etc. This possibility has a natural correlate in colour-theory. The colour-space is continuous and it can be ordered along the three intuitive "dimensions" corresponding to brightness, tone and saturation. Thus it also admits a Cartesian chart: we can devise arbitrary systems for labelling colours by mapping them onto real triplets. But the question of whether or not such a mapping is "correct" or not is undecidable—it is simply a way of describing or indexing the colour-space, and this description is not related to any other phenomena. Only once the colours have been related to other phenomena, for instance the properties of light, by means of measurement procedures can they be said to have an objective significance. And only if they conform to additive properties such as those laid down in Graßmann's axioms do they qualify as *magnitudes.* By operationalising the notion of a spatial magnitude in this manner, Helmholtz was able to finesse the difficulty I just described. Geometry would become an empirical system of measurement that was imposed on a manifold whose elements were not, considered on their own, magnitudes in the full sense of the term. In Kantian language, spatial *quanta* would become *quantities* only once a procedure for measurement was introduced. In a word, Helmholtz applied the representational theory of measurement that was needed to quantify *intensive* magnitudes to the *extensive* manifold of space.

(d) Conclusion

Let us conclude by summarising the theoretical lessons that Helmholtz drew from the three phases I have described above. We need first to characterise the view of geometry, more precisely of spatial magnitudes, that Helmholtz was opposed to already in this early work, in order to determine what he later meant when he argued that geometry was "empirical." As should by now be amply clear, this characterisation cannot amount to claiming that Helmholtz was opposed to a Kantian conception, according to which geometry is *a priori* true, if only for the simple reason that Helmholtz never entirely gives up on the rudiments of Kant's philosophy of science. And in the work we have considered up until now, he employs transcendental arguments quite freely. Thus it would be more correct to say that the empirical geometry he calls for represents an adjustment within Kant's system. It involves strengthening certain of Kant's arguments, while weakening others. Whether this holds true for the papers on geometry from the 1870's, or indeed for still later arguments in *Zählen und Messen*[18] will be reserved for our discussions in the next chapter and in the Conclusion.

Let us then address the first question, namely that concerning the view to which Helmholtz opposed himself in the 1850's. We have essentially two disagreements—the one with Clausius, the other with Graßmann. The first of these consists in denying that purely mathematical magnitudes are admissible in physical theory, all while insisting that

18 (Helmholtz 1887)

certain spatial magnitudes remain legitimate by virtue of their empirical determinacy. The legitimate properties were to include congruence relations, and they were to exclude absolute directions. The second disagreement concerned Graßmann's *a priori* interpretation of his fourth principle. Here, Helmholtz claimed that Graßmann assumed both a given structure of the colour-plane *and* the existence of a measurement procedure which would yield results in conformity with the postulated structure. He objected that both of these could be not simultaneously postulated, unless one was willing to admit definitions of the brightness of different spectral lights that might conflict with our intuitions.

My suggestion is that the problem concealed in Helmholtz's critique of Clausius can be rectified by drawing on the logic of the criticism of Graßmann, and that this is exactly what Helmholtz went on to do when he later resumed his work on geometry. We have discussed the concealed problem a number of times, so I will merely summarise it briefly. In order to make the distinction he needed, Helmholtz needed to claim that congruence relations *were* empirically determinate, even in the case where we are talking about congruent positions of the same system at different points in time, whereas absolute directions were not. This means arguing that the one sort of property (direction) is dependent on mathematical coordinate systems, whereas the other (distance) is not. Helmholtz no doubt believed that his analysis of the notion of congruence in the Königsberger manuscript provided support for this claim: congruence is a motive concept defined with reference to the displacement of rigid bodies, whereas absolute direction would seem to be *a fortiori* empirically indeterminate. But, as became clear in our analysis of the concepts of straight lines and congruence in that manuscript, Helmholtz had actually defined both of these concepts with reference to coordinate-values, and thus the distinction he seeks to draw cannot be made so easily.

What was needed was a definition of congruence that did not involve an appeal to the notions of "length" and "distance." It is here that the colour-research, and Graßmann's vector-calculus opened up new possibilities. For if one operationalises the notion of equal length in terms of rigid-body displacement, the appeal to properties of the coordinate system can be eliminated. One can invert the relation assumed in the Königsberger manuscript, such that distance and length are defined in terms of congruence, as opposed to the other way around. Doing so involves treating rigid bodies and their coincidence-relations as the very basis of the magnitudinal relations among spatial regions. One argues that, just as the colours forming the colour-space have dimensionality and continuity quite apart from any question of their having a magnitudinal structure, physical space has no inherent magnitudinal structure *until* a procedure for determining the equality and additivity of these regions has been given. This can be true all while allowing that the weak topological relations that Kant ascribed to his extensive magnitudes are indeed *a priori* forms of intuition. Helmholtz always insisted that he remained a Kantian in just this sense, entitling the second appendix to the *Facts in Perception*, "Space Can Be Transcendental without the Axioms Being So."[19]

Critics of Helmholtz's methods in the geometry papers have often objected that his use of analytic methods raises problems of its own—Kantian opponents such as Land (the target of the last paper of geometry) objected that these methods are irrelevant to questions concerning the structure of spatial *intuition.* And authors such as Poincaré (who, it is true, does not mention Helmholtz by name) objected that the analytic description itself involves assuming the truth of Euclidean geometry at least at the infinitesimal level. This latter criticism is obviously akin to the first difficulty I identified above in Helmholtz's critique of Clausius. For surely the definition of rigidity involves covert use of Euclidean geometry?

Without wanting to defend Helmholtz's approach in the two 1868 papers on geometry unconditionally—for they are flawed in other respects—I think that we can partly exonerate him if we consider the relation he took to hold between measurement procedures, analytic descriptions, and intuitive spatial relations. The problem with Graßmann's aprioristic interpretation of his fourth additive axiom, as Helmholtz saw it, was that when one combined it with a given measurement procedure, one might get empirical results that were incompatible with the colour-plane one had postulated as valid (in Graßmann's case, a perfect circle). Conversely, if one defined brightness relations "empirically," it might turn out that, on a given measurement procedure, the colours would simply fail to exhibit the required additive characteristics. The colour-continuum might simply fail to be a metric space.

But this possibility alone suggests a transcendental argument that gets one out of the dilemma that Helmholtz had created for himself when insisting that spatial relations had to be empirically determinate. The reasoning connects to the colour controversy quite directly, for there Helm-

19 (Helmholtz 1878a)

holtz had repeatedly confronted the question of whether, and under what conditions, colours might fail to be a metrical space. The lesson he learned was that the colour-space fails to be a space of magnitudes whenever the measurement procedure and the measurement results fail to satisfy Graßmann's definitions of an "additive magnitude." Helmholtz's innovation was to run this reasoning backwards: if some given intuitional space is to satisfy the axioms on additive magnitudes, then it follows that certain kinds of measurements must be possible. And if one can then show on purely analytical grounds that they are possible only under given conditions, then one has transcendental grounds for assuming that these conditions are satisfied. The task would therefore be to demonstrate by purely analytical reasoning that, *if* certain motions of bodies were possible only given a particular metrical structure of space, and *if* furthermore these motions were just those which were required for a determinate system of spatial measurement, then one could conclude that space would have that metrical structure *if* there could be determinate spatial measurements at all.

Such a demonstration would in effect fill in a critical gap in Kant's reasoning. Kant never explains how we could get from the extensive structure of three-dimensional spatial intuition to the specific propositions of geometry. Indeed, Kant does not want to explain this, because the existence of this gap supports the claim that geometrical proofs require constructions, and thus that the propositions of geometry are synthetic. Helmholtz's idea is to show that, among the various extensive structures which are *logically* possible (each of which corresponds to a particular metric), only one is compatible with the regulative demands placed on physical science. These demands are essentially the same that we saw in our discussion of the *Conservation:* it must be possible to comprehend all possible phenomena under position- and time-independent general laws, and the spatial relations that are referred to in these laws must all be empirically determinate.

Thus Helmholtz could respond to both criticisms of his analytic approach by arguing that his formal demonstrations are not directly concerned with the space of *our* intuition, even though it is, in a sense, the properties of the latter which interest us. For in his first two, "transcendental" papers on geometry, the claim is only that if some arbitrary space did not have a particular structure, then certain kinds of operations would be impossible. This demonstration is purely analytical, and thus it in no way depends on the form of our, or any other being's experience. Furthermore, the necessity that our experience conform to the demand that such operations be possible is itself merely regulative, and it could in fact be frustrated. And so what Helmholtz has shown is that if our experience is to be completely comprehensible, then it must exhibit at least as much regularity as is required to define fundamental spatial magnitudes. The formal demonstration does not depend in any sense on what we do or do not intuit. And his response to the more sophisticated, conventionalist objection, would take the same form: in order to characterise the structure of the various manifolds under consideration, one has to make use of algebraic propositions that seem to concern the space of our intuition. But since their truth is dependent on numerical relations (which Helmholtz considers to be analytic), there is no need to suppose that. Furthermore, the question of *whether or not* such numerical relations are applicable to our experience will in fact be settled only once we actually go about the work of empirical measurement.

The role of regulative laws in this theory brings us to the last of the topics discussed in the previous chapters which shall be relevant to our discussion in the last two. This concerns the notion of a complete determination of the system of nature, which we discussed at length in Chapters 2 and 3. We should recall that the entire logic of the criticism of Clausius depended essentially on a transcendental argument: non-central forces may be mathematically possible, but they are empirically indeterminate. Insistence on empirical determinacy is itself justified with reference to the regulative goal of completely grasping, that is to say of giving a complete determination of nature. There is, once again, no constitutive guarantee that this is possible. But suppose one accepts this regulative goal, and suppose one also claims, as both Kant and Helmholtz do, that a complete determination of nature must take the form of general mathematical laws. Then one can extend the transcendental argumentation to the level of the spatial magnitudes themselves. Roughly, if there are to be general physical laws governing the behaviour of bodies at all points in space and at all times, then these parts of space must be quantitatively comparable with one another. But they would not be so if bodies could not be displaced in such a way as to satisfy certain axioms. Thus the complete comprehensibility of nature entails that such bodies exist—not in the sense that the contrary is ontologically impossible, nor even unimaginable, but in the far weaker sense that we will simply fail in our regulative aims if they do not. In the next chapter, we shall see how this regulative demand was itself reformulated as Helmholtz moved from his initial, and flawed transcendental demonstration of the validity of Euclidean geometry to a more radical position.

6. Helmholtz on Geometry, 1868–1878

In the period between 1855 and 1868, when he published the first two of his seminal papers on geometry, Helmholtz was named professor of physiology in Heidelberg (1857), and devoted himself primarily to research in sense-physiology. This research, which culminated in the publication of two monumental works, the *Handbook of Physiological Optics* (1860–1867) and *On the Sensations of Tone* (1863), occupied most of his time in this period, and he accordingly did little research in physics and mathematics. In 1868, however, he was offered a chair in physics at the University of Bonn, at which point he was already gearing up for his return to the field he had always viewed as his calling. The period immediately prior to 1870, when he was offered a chair in Berlin, saw a burst of intensive activity in the field of electrodynamics, the results of which work were published in a series of papers beginning in 1871. Given the wide scope of his activity at this time, it will evidently not be possible within the confines of this study to track his development in all these fields. What is important to note, however, is that his investigations into the metrological foundations of geometry coincided with his return to properly physical investigations, in particular to electrodynamics. If I am right in identifying the importance of metrological considerations in the Kantian arguments of the *Conservation*, and above all Helmholtz's response to Clausius's attack, then this should not come as a surprise. Furthermore, this correlation would speak against against the view that this research is primarily derived from concerns emerging within his physiological research on perception.[1]

Such an approach to Helmholtz's work on geometry seems to me misleading on two counts. First, although there can be no doubt that Helmholtz adhered to a version of Kant's epistemology that had been naturalised at the hands of Herbart, Wundt and Lotze, it remains the case that he presented arguments for the empirical character of geometry that had a specific philosophical and mathematical content. This content did not stop with the claim that spatial intuition had a physiological basis (and

1 This reading was made as early as (Land 1877), who describes Helmholtz as "fresh from the physiology of the senses," p. 40.

thus could have been different)—a claim which Kant *might* have been prepared to accommodate, but which *could* in any case be made consistent with a transcendental epistemology—rather it related directly to the role of *measurement.* And this points to the second shortcoming of an overly physiological reading of these works—a shortcoming that is indeed shared by later interpretations of Helmholtz's philosophy, such as those offered by members of the Vienna Circle. If one reads Helmholtz primarily as advancing a theory concerning the way in which spatial data are organised by the developing epistemological subject—that is, if one focusses on those portions of his discussion which concern what we would call today developmental psychology—one may very easily come to think that he did not consider adequately the theoretical role of geometry in physics. Concretely, one may be led to ignore those passages in the geometry papers where Helmholtz emphasises the connection between measurement by means of congruent bodies and the role of spatial magnitudes in Newtonian science.

For instance, in "Über den Ursprung und Sinn der geometrischen Sätze,"[2] the last of the papers, Helmholtz defines what he calls "physically equivalent magnitudes" as those spatial magnitudes in which the same physical processes take place in the same period of time. He then observes that the most common method for determining these distances is by means of congruence determinations with rigid bodes. Commenting on this passage, Schlick suggests that the first definition is an aberration, and goes on to castigate Helmholtz for overlooking the possibility of a conventionalist interpretation of geometry.[3] Without going into this controversy in greater detail at this point (I will return to it at the end of this chapter), we can summarise the thrust of such objections as follows. Helmholtz was arguing that the propositions of geometry were inductive in origin, and he did so within a psycho-physiological framework. Because of this, he overlooked the role of geometry in mathematical physics, where it is always coupled to laws of motion. His blindness in the face of the conventionalist interpretation derives from this overly physiological bent. For he fails to see that what we come to regard, on inductive grounds, as equal distances in space, must be subject to correction when we do mathematical physics, and that this correction will ultimately be undertaken with reference to geometric *norms.*

2 (Helmholtz 1878c)
3 (Helmholtz 1977), p. 183. *Cf.* footnote 45 of this chapter.

Now, I am obviously not opposed to reading Helmholtz from the point of view of sense-physiology. I would, however, urge that we look to Helmholtz's scientific research in that domain, and not to his popular lectures if we want to understand what the contribution of the sense-physiological research in fact was. The problem with using the papers on geometry as a point of departure for our investigation is that they are disparate works aimed at disparate audiences. This would in itself not present a problem but for the fact that the third and fourth papers, which contain the bulk of the philosophical arguments that have received scrutiny, are written for the least sophisticated readers. This cannot license us to ignore those arguments, any more than Schlick is licensed to do so. It does mean, however, that the choice of examples used by Helmholtz, along with the greater emphasis he lays in these papers on the possibility of visualising non-Euclidean spaces, are a poor indicator of Helmholtz's actual views. These examples are intended to counter a specific criticism raised by Kantians such as J.P.N. Land. On this objection, even though the *analytic description* of non-Euclidean geometry is logically permissible, it remains the case that case that we cannot *intuitively imagine* such geometries. Helmholtz's response to this objection is two-fold: first, he offers a series of thought-experiments in which we *can* imagine non-Euclidean relations among material bodies; second, he argues that—one way or the other—spatial relations which are not empirically determined are of no *utility.*

It is the second of these responses that carries the greater argumentative burden, and it is one which, as we have seen in detail already, goes back to Helmholtz's earliest work on the philosophy of physics. Whereas the criticism of his use of analytic methods, along with Helmholtz's reply that non-Euclidean geometries are indeed visualisable, is, by contrast, a local objection to the methods he used to carry out the argument from empirical determinacy. Thus Helmholtz gives a series of recipes for constructing non-Euclidean intuitions, and he suggests that had we grown up in a world in which such intuitions predominated, we would have opted for a different geometrical system. Here again, we must exercise care in interpreting the philosophical significance of this developmental picture. It is true enough that Helmholtz systematically conflates the inductive procedures of an ideal infant with those of the working scientist: on his view the process of induction starts in the cradle, and science is merely an extension of such embryonic inductive methods later in life. And there can be no doubt that this fusion of child development with philosophy of science—a common enough conceit in the history of philosophy—sends an important philosophical message. As Hatfield, Turner and others correctly argue, Helmholtz is using the results of his geometrical investigations as a cudgel against nativism.[4] If these results are correct, then nativism must be wrong. This may even explain in part why he undertook these investigations. But quite evidently he cannot have thought that the proof of these arguments depended on assuming that nativism was wrong. On the contrary, he had to provide an independent justification for their validity. His final position is therefore: (1) it is logically possible that we would have evolved non-Euclidean intuitions of spatial relations; however, (2) whether our preferred intuitive geometry was Euclidean or not, the only empirically relevant geometry would be that required by physics for the purpose of establishing general laws of nature. The latter point remains, as it was all along, the decisive one.

(a) The Four Papers

Helmholtz's papers on geometry can be divided into two groups: the first two, entitled "Über die tatsächlichen Grundlagen der Geometrie" and "Über die Tatsachen, die der Geometrie zum Grunde Liegen" were published in quick succession in 1868.[5] The first, a report to the *Naturhistorischer-medizinischer Verein* in Heidelberg, is basically a summary of the main lines of the mathematical argument presented in the second. It does, however, contain some philosophical material absent from the second, more technical article. There has been repeated confusion in the literature due to a misdating of the first of these papers. In the *Wissenschaftliche Abhandlungen*, the shorter report is dated 1866, and this has led some commentators to see a divergence in views and methods, which is then ascribed to Helmholtz's gaining access to Riemann's "Über die Hypothesen, welche der Geometrie zu Grunde liegen"[6] in the intervening period. I will treat the two papers as a unit in the first section of this chapter.

The next two papers, "Über den Ursprung und Bedeutung der geometrischen Axiome"and "Über den Ursprung und Sinn der geometrischen Sätze; Antwort gegen Herrn Professor Land"[7] are similar in their

4 (Hatfield 1990; Turner 1994)
5 (Helmholtz 1868a, 1868b)
6 (Riemann 1854)
7 (Helmholtz 1870, 1878c)

arguments and rhetoric, but have quite different publication histories. The first of them was a lecture delivered only two years after the mathematical papers to the *Docentenverein* in Heidelberg, that is to say to an audience of *Gymnasium* teachers. Helmholtz makes no bones about the fact that he is speaking to laymen, and that his presentation of the topic is in some measure superficial. Nevertheless, the paper is not simply a popular version of the first two. First of all, Helmholtz had become aware of work by Beltrami which revealed that his first two papers contained a significant error: he thought he had proved that the validity of Euclidean geometry followed from a series of assumptions concerning the possible displacements of rigid bodies in an unbounded, continuous manifold; however, as Beltrami's work showed, his assumptions were consistent with pseudo-spherical geometries. Far from seeing this as a setback, Helmholtz took this correction to provide more conclusive evidence for his earlier claim that geometry was an empirical science, and he sought to exploit this evidence in the 1870 paper. The second important difference between this paper and the earlier two is that Helmholtz talks here about the relation of geometry to physics, and thereby explicitly contrasts his views to what he calls a "strict Kantian" insistence on the *a priori* nature of Euclidean geometry. These passages are of great significance, and will be discussed in section two of this chapter.

The last paper was originally published in English under the title "The Origin and Meaning of Geometrical Axioms (II)" in *Mind.*[8] The paper was a reply to a series of objections that had been raised by J.P.N. Land in an article in the same journal from the previous year,[9] and which referred in turn to the 1870 paper, which *Mind* had also published in an English translation.[10] Land chides the scientist, "fresh from the physiology of the senses,"[11] for his inability to distinguish between abstract and analytical geometries on the one hand, and the geometry of spatial intuition, which determines the form of all possible objects that we can represent to ourselves in our non-scientific, phenomenal experience. Land's argument boils down to a clear distinction drawn towards the end of his paper, and which was evidently the picture that Helmholtz responded to. Adopting Helmholtz's example from the

8 (Helmholtz 1878b)
9 (Land 1877)
10 (Helmholtz 1876)
11 (Land 1877), p. 40.

1870 paper of two-dimensional beings living on the surface of a sphere, Land imagines that,

> ... some genius among them might conceive the bold hypothesis of a third dimension, and demonstrate that actual observations are perfectly explained by it. Henceforth there would be a double set of geometrical axioms; one the same as ours, belonging to science, and another resulting from experience in [*sic.*] a spherical surface only, belonging to daily life. The latter would express the 'object' of sense-intuition; the former, 'reality,' incapable of being represented in empirical space, but perfectly capable of being thought of and admitted by the learned as real, albeit different from the space inhabited.[12]

For these beings, in other words, spherical geometry is the geometry of intuitive experience, and Euclidean geometry is the geometry imagined by scientists and mathematicians to hold of the "real world." Kant's doctrine of spatial intuition concerns only the first of these, and thus it is immune to the purely analytic speculations of the scientist, even if the latter lay claim to some higher, metaphysical validity for their imagined geometries. Land also objects, in terms that certainly would have found approval with Kant, to Helmholtz's use of the notion of a rigid body in his definition of congruence. Such definitions are, on Land's view, perfectly permissible, but they would belong properly to phoronomy, since they involve the notion of matter, whereas geometry considers only relations among parts of space. This analysis of the situation is quite accurate, as we have already had occasion to see. For Helmholtz had from the beginning insisted that the notion of congruence, although it was a "general concept of nature," was a pure and *a priori* motive concept, which characterisation agrees quite well to Kant's definition of phoronomy.

Thus the last of Helmholtz's geometry papers, which is in many respects the most reflective from a philosophical point of view, is also a direct response to arguments of an orthodox Kantian opponent. We shall therefore use it, in the last section of this chapter, to tie together the earlier discussions of Kant and the arguments of the *Conservation* to the specific innovations of the first three papers on geometry. In this last paper, Helmholtz tacitly accepted Land's characterisation of his arguments as phoronomic, even if the rhetoric of his reply is, as always, uncompromising. He responded to the first of Land's criticisms—namely that the possibilities introduced by analytic methods are more numerous than those admitted by the form of spatial intuition—by restricting his earlier visualisation examples to manifolds of lesser dimensions. But in doing so,

12 (Land 1877), p. 44.

he explicitly maintained that the geometrical figures that one was to imagine were material. This insistence was already evident in the early Königsberger manuscript, and recurred in the reply to Clausius as the demand that spatial determinations be conceived as relations among real things. The application of physics to the material world presupposes that we can *realise* the definitions of its empirical units. I use the term "realise" here in the sense that it is used in modern metrology. Here one distinguishes between the theoretical *definition* of a unit (the meter is a unit of length connected to other basic units by means of the fundamental laws in which they appear), its *representation* (it is a certain number of wavelengths of a given light frequency) and its *realisation* (an actual experimental system that is stipulated to agree to the representational definition). Our units of length are realised by sets of rigid bodies, and thus the behaviours of these bodies when they are displaced are the determinants of space that we codify by means of geometrical propositions. According to Helmholtz, the properly transcendental properties of space are thus to be restricted to dimensionality and continuity, while its metrical properties depend on motive characteristics of the bodies used to determine those congruence relations which are relevant to physics.[13] This terminal position explains why Helmholtz argues, in this last paper, that the goal of geometry is to establish "physically equivalent magnitudes," in which the same processes occur in equal periods of time, thereby explicitly connecting geometry and physical laws in a science indistinguishable from Kant's phoronomy.

(b) The Papers of 1868

We can only speculate on Helmholtz's reasons for recommencing his work on the foundations of geometry at the end of the 1860's. But there can be little doubt that they derived in part from his imminent return to full-time physical research. And we know that the work on colour-theory, as well as on the displacement of retinal images, had occa-

13 Even though he calls the specific axioms of geometry into question, Helmholtz remains open, if not committed, to an essentially Kantian theory of spatial intuition. This is reflected in the title to the second appendix to the *Facts in Perception* (Helmholtz 1878a), "Der Raum kann transzendental sein, ohne daß es die Axiome sind." The 1878 geometry article was first printed in German in abbreviated form as the third appendix.

sioned mathematical investigations of two distinct sense-physiological domains that were relevant to his understanding of the problem.[14] The first of these has been discussed in detail in Chapters 4 and 5, so that we will have little more to say about it here. Parallel work on eye motion and retinal images aimed to give an explanation of Donders' Law, which stated that to each direction of the principal axis of the eye (to each direction in which one can look directly) there corresponded a unique angle of torsion, i. e. a unique degree of rotation of the eyeball about the axis determined by this line of sight. Helmholtz explained this phenomenon by arguing that it was obviously beneficial if the image projected on the retina by an object retain a rigid connection (*einen festen Zusammenhang*) among its points as the centre of vision was moved about the object. This criterion was easily fulfilled when the motions were slight, or when they lay along "meridians" passing through the primary position (that in which the eye looks straight ahead). In the case of large motions that did not follow meridians, however, it could not be met, thus some distortion was inevitable. The angles of torsion which Donders' Law predicts are precisely those, Helmholtz argued, which result in a minimum of such distortions when the eye sweeps the visual field. Simply put, we do our best to preserve the relations holding between the points of the image on our retina as we scan.

It is not hard to see how this research connects up to one major strand of the two papers of 1868. Here, Helmholtz sets out to define the conditions under which measurements in space can be carried out at all. Since, he argues, measurements depend essentially on our being able to establish relationships of congruence between bodies in space, we must postulate that these bodies preserve rigid connections between their various points independently of their locations and orientations. This property was not fulfilled in the case of the visual field, for an after-image (corresponding here to a displaced rigid body, or measuring instrument) which under some motions would appear congruent to another image in the visual field, would fail to be so in others. This difference depended on the initial position of the eye, as well as on the path it had to take to get to its final one. Visual afterimages do not, in other words, fulfil the condition of free mobility that Helmholtz set on his rigid bodies:

> This means in other words that the congruence of two spatial images is not dependent on their situations, or, all parts of space are congruent to one an-

14 *Cf.* (Helmholtz 1863a, 1863b), which Helmholtz refers to at (Helmholtz 1868b), p. 619. These papers are discussed in (Lenoir 1993).

> other if one disregards their delimitation [*wenn von ihrer Begrenzung abgesehen wird*], just as all parts of the same spherical surface are congruent with regard to the curvature of the surface, if one disregards their delimitation.
>
> The visual field displays a more limited mobility of retinal images on the retina. I have dealt with the particular consequences which ensure for the estimation of distances by means of the eye [*Augenmaß*] in my physiological optics.[15]

There are indeed notable similarities among Helmholtz's mathematical methods in "Über die Tatsachen, die der Geometrie zum Grunde Liegen" and "Über die normalen Bewegungen des menschlichen Auges." In both cases, Helmholtz postulates that internal relations between a system of points remain constant (in the geometrical case) or are held to a minimum (in the optical one) as the system undergoes infinitesimal displacements. In both papers, he goes on to derive expressions for all possible transformations of the coordinates of the system that are consistent with his respective postulates.

Taken together, the work on eye motion and the work on colours were valuable resources that Helmholtz could draw on in his new geometrical investigations. But the contributions of the two were nonetheless dissimilar: whereas the work on eye motion provided largely mathematical resources, the colour research had isolated the role of measurement operations in determining the metric of a manifold. Furthermore, Helmholtz had been challenged in that research programme by competitors whose analysis of the problem had in many respects surpassed his own. Even though Helmholtz gave Graßmann and Maxwell short shrift (his admiration for Graßmann was unfortunately only expressed once the latter had passed away), Graßmann's mathematical analysis and Maxwell's quick application of the latter were better than what Helmholtz could come up with at the time. Leaving questions of priority aside, however, it remains the case that Graßmann and Maxwell forced Helmholtz to think about what it would mean to say that distance relations in an intuitive manifold were quantitatively defined. This was important precisely because Helmholtz had already used the appeal to quantitative determinacy in order to argue against what I have been calling "absolute" conceptions of spatial magnitudes. And Helmholtz thought that certain electromagnetic theories that disagreed with the *Conservation* programme could be invalidated on the strength of their presuming such absolute magnitudes. Thus showing that the science of spatial measurement

15 (Helmholtz 1868b), p. 624. *Cf.* (DiSalle 1993), p. 511.

could be well-founded without ascribing an inherent metrical structure to space (to put the matter slightly anachronistically) was, on my reading, a matter of tactical importance to him.

The methodology of the 1868 papers was at root the same as that of the Königsberger manuscript on general physical concepts. Helmholtz uses analytical methods to describe the properties of our intuitions of space that are required if the latter is to be adequate to "containing" bodies. Nevertheless, the problem being addressed has been whittled down considerably. The early manuscript considered both the conditions on the containment of bodies that are not in motion (thus, dimensionality, continuity and distance), as well as those holding on the representation of bodies that are displaced (rigidity and thus congruence, straight lines and thus inertial paths). In these later papers, by contrast, Helmholtz is concerned with only one property, namely congruence. No explicit distinction is made between motive and non-motive properties of space. Furthermore, the possibility of determining congruence is now accorded a fundamental role:

> My point of departure was that all basic [*ursprüngliche*] spatial measurement depends on the determination [*Constatierung*] of congruence, and thus that the system of spatial measurement must presuppose those conditions under which alone we may speak of determining congruence.[16]

Thus Helmholtz makes clear from the outset that his method departs from a regulative demand: we require that spatial measurements be made, and such measurements presuppose the possibility of determining congruence. The question is then to determine *what exactly* the realisation of this possibility entails. Helmholtz's answer, at the end of the second article, is that the "possibility of the system of our spatial measurements ... depends on the existence of natural bodies that agree adequately to the concept of rigid bodies that we have specified [*i.e.* to the postulates Helmholtz sets up in the paper]."[17]

This appeal to regulative conditions on measurement is not, as I have argued in the previous chapters, a Kantian anomaly in the argumentation of an ardent empiricist. On the contrary, it is a perfectly natural step in a chain of reasoning that began before the *Conservation,* and which culminated in the demand, in the reply to Clausius, that one not appeal to purely mathematical coordinate systems when doing physics. Because

16 (Helmholtz 1868a), p. 614.
17 (Helmholtz 1868b), p. 639.

the task of natural science is to set up general laws, and because these laws are to be mathematical equations whose terms ultimately resolve onto basic spatial and temporal magnitudes, these latter magnitudes must be determinate. But empty space is not determinate. Thus the magnitudes in question must be determined by real things in space. Nevertheless, Helmholtz's statement of the problem is far more precise than what he had been able to give previously. As he now sees it, the ultimate reduction of statements about spatial magnitudes issues either in a loop, or in an absurdity: suppose that we stipulate that all statements about the lengths of a set of bodies B are to be arrived at by measuring them with a set of bodies M according to certain rules. Both the members of B and M are physical bodies, and so we know in advance that the characteristics of the measurement bodies M cannot agree precisely to the axioms of geometry. Conversely, in order to say whether or not the M's are better or worse measuring instruments, we investigate the degree to which they do fulfil the axioms:

> … the spatial forms of geometry are ideals to which the physical forms of the real world can only approximate, without ever completely satisfying the demands of the concept [i.e., of our geometrical concepts], and because we must test the invariance of form, the rightness [*Richtigkeit*] of the planes and straight lines that we discern on bodies, precisely by means of these same geometric propositions….[18]

The loop arises from the fact that, so long as we perform actual measurements, all comparisons are between imperfect bodies that do not satisfy our ideal demands. To escape this loop would involve the absurdity of comparing a physical object with a mathematical ideal. In actual practice, we can avoid this dilemma quite easily: we don't need a fully articulated theory of our measuring instruments in order to go about measuring quite successfully, and so we can remain content with comparative measurements of the first sort. But this is not an option when we are investigating the fundamental conditions on spatial measurement. For such an investigation would seem to demand that we evaluate propositions expressing geometric truths with reference to these same truths, which is circular.

This dilemma therefore forces us, on Helmholtz's account, to turn to analytical methods. They permit us to describe the conditions under which measurement could, or could not, take place in conformity with the axiomatic demands we place on it. The propositions of analytic ge-

18 (Helmholtz 1868a), p. 610.

ometry do not depend for their validity on the existence of the structures they describe, for analytic geometry "calculates with pure magnitudinal concepts, and requires no intuition for its proofs."[19] Helmholtz is in other words making a sharp distinction between geometry in the traditional sense, and analytic geometry. The former is a science of spatial measurement: its objects are very much physical objects, but these objects are considered by geometry only with regard to their extension. Analytic geometry, on the contrary, is not concerned with real things at all, so that its truths hold of any object that can fall under its concepts, whether or not these are realised physically, or for that matter in pure intuition. Helmholtz has thereby already ruled "pure" geometrical intuitions out of court, although he does not explain this further here (he returns to this point in the 1878 reply to Land). Since *intuited* spatial magnitudes cannot be compared to physical objects directly, they cannot play a role in measurement. And since the relations that hold among them can be described by means of another, more general discipline (physical geometry), they also have no epistemological role to play.

The mathematical set-up of "Über die Tatsachen" (the longer, mathematical paper)[20] involves a specification both of a three-dimensional continuous manifold with a local Cartesian chart, as well as of the rigid bodies that are presumed to exist in this manifold. Their rigidity is then defined by means of an equation holding between their material points. This equation is independent of their motion (that is to say, of changes in their coordinate values), and it thereby defines classes of congruent bodies: two bodies are congruent if and only if they have the same number of points, and the latter are related to one another by the same equation, for then they can always be made to coincide with one another.

These conditions do not yet fulfil all demands that we place on our measuring instruments, however. In order for every spatial distance to be comparable with every other, we must presuppose that the rigid bodies can be arbitrarily transported (free mobility), and that they return to the same position when rotated (monodromy). Helmholtz then goes onto show that it follows from these postulates that the equation characterising the relation between two points defining a (minimal) rigid body must be of a generalised quadratic form, namely it must have what we

19 (Helmholtz 1868a), p. 611.
20 (Helmholtz 1868b)

now call a Riemannian metric.[21] This defined the class of possible metrics as those of constant curvature, which Helmholtz erroneously thought to be restricted to spaces of zero or positive curvature, that is to Euclidean or spherical geometries. By adding the postulate that the space in question was unbounded, he was able to eliminate the spherical geometries from consideration, and thereby arrived at the mistaken conclusion that Euclidean geometry was uniquely selected by his postulates.

Thus Helmholtz had shown, as advertised, that there would be no spatial measurements at all (or at least none that meet the general demands we place on measurement), without the possible existence of certain regularities in the behaviours of bodies. He summarises these results as follows:

> At the same time I call attention to the fact that the possibility of our system of spatial measurements, as has been clearly demonstrated in this treatment, depends on the existence of such natural bodies as agree to the definition of a rigid body that we have supposed. The independence of congruence from positions, from the orientation of coincident spatial forms, and from the path that they follow when brought together, is the fact on which the measurement of space is based.[22]

It is striking that in this concluding passage, Helmholtz makes no use of the word "geometry," which omission would seem to weaken his thesis. For if Helmholtz is merely saying that there can be no measurements unless there are rulers, then we can hardly disagree with him. Whereas, as he himself emphasises in the opening sections of the same paper, there is somewhat more at stake here. What gets to count as a ruler is judged with reference to geometrical propositions. We demand a certain regular behaviour of our measuring instruments, and the norms according to which those behaviours are evaluated are just those that are codified in geometrical propositions. Helmholtz seems guilty of a bait and switch: What we were promised was a demonstration of the fact that certain of these geometric *norms* "express truths with factual content,"[23] whereas what we are given in conclusion is a demonstration that if there were no bodies satisfying geometric propositions, then there would be no measurement. How are these two propositions related in Helmholtz's concep-

21 *Cf.* (DiSalle 1993) for a summary of Helmholtz's argument, as well as references to the work of Lie, Russell and others.

22 (Helmholtz 1868b), p. 639.

23 (Helmholtz 1868b), p. 639.

tion of the problem? Indeed, how could we ever claim that a *norm* was a factual truth?

Here we must recall the transcendental argumentation of the *Conservation*, where Helmholtz also appealed to his principle of positional determinacy in order to derive propositions concerning possible forces. There, Helmholtz applied what I called the "downward" determinacy requirement in order to show that the actions of forces are central. The argument ran:

- forces are causes of changes in the relation between empirically given bodies;
- a pair of bodies has only one relational property, namely the distance separating the pair;
- thus forces can only cause changes of this magnitude;
- thus the action of a force is directed along the straight line connecting the bodies.

Similarly, he applied the "upward" determinacy requirement to show that if there were changes in the actions of a force, these changes had to be functions of the distance. Failure to meet this requirement would mean that such forces were not subsumable under higher and invariant laws. This would violate the regulative demand that nature be completely comprehensible.

I contend that both the regulative and constitutive (upward and downward) requirements are at work in the 1868 papers, and indeed that the argument of these papers is scarcely comprehensible without that context. The regulative, upward part stems from the demands we place on geometry. We require a system of spatial measurement in order to do physics.[24] This system must conform to certain demands. It must be a *realised* system, in that it consists not merely of definitions, but of the actual empirical systems that are referred to in these definitions. Thus it presupposes the existence of bodies that can be used to carry out measurements. Furthermore, it must permit the comparison of arbitrary spatial distances. From this requirement, we can derive the postulates of free mobility and monodromy. All these demands are evi-

24 This is clearly stated at the opening of (Helmholtz 1870), p. 2. "Land-surveying and architecture, mechanical engineering just as well as mathematical physics—they constantly calculate spatial relations of the most diverse kind according to geometric propositions. They expect that the results of their inventions and experiments will conform to these calculations...."

dently norms, for they stipulate what has to, or ought to be, if we are to carry through to our goals. The constitutive demands derive from properties of the spatial manifold itself: continuity, dimensionality, perhaps unboundedness. These properties place minimal bounds on possible empirical intuitions, some of which, in turn, must conform to the regulative demands. Put simply, we must assume for regulative reasons that we live in a world in which rulers could exist.

Despite the similarity in the approach employed in the *Conservation* and that used in the 1868 geometry papers, the latter investigation must employ a somewhat different methodology. The problem is not so much with the regulative demands, for they are well-defined, but with the constitutive ones. The definition of possible forces assumed that we could speak of spatial (and temporal) quantities without running into excessive difficulties. Helmholtz disallowed absolute coordinate systems and absolute directions, but he did not actually problematise the quantitative relations holding between regions of space so long as these were empirically delimited. He did demand of Clausius that the coordinate systems used to cash out his conservation principle (to say that a system was in the same internal state at two points in time) could not be arbitrary, but he apparently assumed that once the points in question were empirically given, the magnitudes would be fixed, and thus the metrical relations as well. Similarly, in the Königsberger manuscript, he had defined congruence in terms of equal length, which again involved assuming that metrical properties are properties of the spatial manifold that we can simply invoke. But in the geometric case, he was calling just this proposition into question. His refusal to permit transcendent determination with respect to absolute space, when pushed to the extreme, forces us to specify the constitutive conditions with entirely different means. The difficulty is at root the same one Helmholtz raises when talking about the ideality of geometric objects: ideal objects cannot be compared with physical objects; furthermore, so long as they are consistent, the propositions of geometry cannot be used to check their own validity without begging the question. In this more technical setting, the problem is better defined. In order to talk about the conditions for determining metrical relations, we have to assume, at least provisionally, that space has such relations (i. e. that it is, at the very least, locally Euclidean); however, since we are supposing that these are not empirically accessible to us, they can be of no help to our characterisation. How can we talk about the properties of space at all without assuming that it has *some* metrical properties?

The advantage of algebraic methods is then evident. Since the task at hand is to define the conditions under which measurement is possible, we need to be able to describe conditions under which it *wouldn't be* possible. We can't do this so long as we stay within the system of geometrical propositions, since the best we can do in that case is check consistency.[25] But because analytic geometry is a completely abstract science of magnitudes, it does not restrict us to considering those manifolds in which measurements could actually be carried out by an epistemological agent. By employing an analytic description, we can distinguish between worlds in which measurement is possible in the required sense, and worlds in which it is not. The distinguishing characteristics of the two classes of world will then be the conditions that we are looking for.

It is just in this regard that Helmholtz's method differs from Riemann's.[26] Riemann's work represents what Helmholtz had called a "purely mathematical" investigation of possible manifolds. After he has introduced the notion of a continuous manifold with a local Cartesian coordinate system, Riemann defines the distance on the manifold in its general form. In contrast, Helmholtz investigates what we might call manifold/operation pairs, much on the model that he learned during his colour-research. There, as we may recall, the metrical relations on the colour plane were partly the result of additive axioms and a stipulated measurement procedure, and partly the result of the actual measurements one carried out. The colours had no inherent quantitative relations until they were related to an external domain by means of measurement procedures. Similarly, a Helmholtzian geo-*metrical* space is a manifold plus a concrete measurement operation and its attendant axioms. Together, the measurement operation and the axioms define the magnitudes on the manifold, just as they defined colours as magnitudes. These magnitudes are—by design—not purely mathematical, but ones that are physically realised.[27] Furthermore, they satisfy all the regulative demands we place on measurement, for these are embodied in the axioms of free mobility, monodromy, etc. Because he overlooked the compossibility of Euclidean and pseudospherical geometries, Helmholtz argued that only Euclidean geometry

25 We cannot show consistency in the sense that we do so in a meta-logical proof. But we could, if the axioms were inconsistent, show that they were so.

26 (Riemann 1854)

27 Again, I use the term realised in the sense of modern metrology—a definition is *realised* once an actual device for measuring is correlated with the definition (or, strictly speaking, with its "representation").

was consistent with these regulative and constitutive demands. Thus only Euclidean geometry satisfies the dual demands that all regions of space be comparable, and that all spatial magnitudes be conceived as relations between real things.

So, in response to our question from above (How does the requirement that such bodies exist show that the propositions of geometry have a factual content?), the 1868 papers suggest the following answer: the constitutive *a priori* properties of spatial intuition (its topological properties) do not determine a metric; but when we introduce those assumptions concerning the existence of bodies that would make the determination of metrical relations possible, it ensues that the metric must be Euclidean. Thus Euclidean geometry is true—in the sense of usable—only if such bodies exist. That they exist is presupposed by the sciences, but just as with every other regulative demand, we could be disappointed in this postulate. It is regulatively necessary, but it remains for all that an empirical fact.

These considerations can now be used to answer our question from above concerning the normative role of geometry. I suggested that Helmholtz apparently does a bait and switch: we are to be shown that geometrical propositions, which we use to check the suitability of measuring instruments for their task, have an empirical content. Thus a norm will be shown to be an empirical fact. But it would seem that at the end of the investigation, we have been shown that there would be no measurement without measuring instruments—a proposition that no one would call into doubt. So what happened to the normative component? Surely Helmholtz has changed the subject.

Nonetheless, I do not think that Helmholtz is guilty as charged. We must keep in mind the structure of the transcendental argument to see why. The normative aspect of geometrical propositions derives from the regulative *demands* placed on measurement. We evaluate any possible system of measurement by investigating the degree to which it fulfils our normative requirements. But, just as in the case of the construction of forces, these regulative demands are not all there is: we might regulatively demand that it be possible to make statements about the magnitudinal relations holding between distinct regions of space without having the means at our disposal to do so. Indeed this is just what Helmholtz thinks we do when we interpret geometry as a synthetic *a priori* science in Kant's sense. We claim implicitly that certain comparisons can in fact be carried out. Making this claim explicit means spelling out how these comparisons are to be realised with real objects. Such procedures, like the procedures for adding colour, are conditions on the meaningfulness of quantitative comparisons. Since Helmholtz believes himself to have shown that such comparisons are only possible on the assumption that bodies exist which conform to the regulative demands, the *empirical meaning* of quantitative spatial comparisons depends on contingent states of affairs—*Tatsachen* or facts, as Helmholtz calls them. The normative force of geometry derives from our need to measure spatial relations. Whether or not we can satisfy that need remains contingent. But that is of course the case with all regulative demands.

I should emphasise that, although all the components of this theory are essentially Kantian, it nonetheless ascribes a completely different epistemological status to geometry than what we find in Kant. The model that Helmholtz is working with is the one that he had already developed in the *Conservation*—what I called "squeezing" the necessary form of concepts between regulative and constitutive demands. The regulative requirement of the comprehensibility of nature, which enjoins us to construct general laws, requires that these laws subsume spatial and temporal relations among material phenomena. Thus natural science assumes that such spatio-temporal magnitudes are generally available, and that they can be empirically determinate. By adding in constitutive constraints (such as that two points define only a single spatial magnitude), we can then draw specific conclusions concerning the possible form of physical concepts. In this, Helmholtz's methodology was extremely close to Kant's, at least when it was a matter of determining the transcendental requirements placed on the concept of force. But Kant's theory of geometry does *not* use this approach: geometric propositions do not derive in any sense from regulative requirements, for they are arrived at by constructing the pure intuitions that correspond to the "mathematical" categories.

Furthermore, both Kant and Helmholtz employ this method within the context of material science, albeit the pure metaphysical part thereof. The constructions in question are determinations of empirical concepts, meaning that the latter subsume empirical data, even if these are, once again "pure empirical" ones. The same goes for Helmholtz's method in the 1868 papers on geometry. Thus we can agree with Land that what Helmholtz is doing is, from the point of view of Kant's system, a kind of phoronomy. Indeed, I suspect that Helmholtz too would have agreed with this characterisation. In the Königsberger manuscript, congruence relations are defined in terms of the displacement of rigid, mathematical bodies. Here, in the 1868 papers, the demand is strengthened and made explicit: these rigid bodies are not idealities, they must be empirically given. Finally, since they are considered independently of, and prior to, any dynamical concepts, they correspond quite exactly to the objects of Kant's phoronomy, which considers movable *matter* in space without regard to the category of causation.

(c) "Über den Ursprung und Bedeutung der geometrischen Axiome" (1870)

Shortly after the two 1868 papers were published, Helmholtz became aware of work by Beltrami that showed he had overlooked pseudo-spherical geometries. This error was rectified in footnotes and an appendix when the papers were later published in the *Wissenschaftliche Abhandlungen.* But he endorsed this corrective immediately in his public lecture of 1870, indeed he took it to strengthen the empiricist line that he had been urging in the original papers. Whereas in the deduction we have just outlined, Euclidean geometry is shown to be a transcendentally required system of empirical truths, it now results that the transcendental requirements placed on measurement do not single out a unique metric, and thus they leave open the question of which geometry is valid for our experience. Helmholtz took this to show that the decision between pseudo-spherical and Euclidean geometries depended on experience. But in taking this step, he also acknowledged that other responses were possible: since either of the two geometries could satisfy the requirements placed on measurement, it would also be possible to *stipulate* that the one or the other was correct, and to adjust the other laws of physics accordingly. This option is that later endorsed by Poincaré, and it corresponds roughly to what we call the *conventionalist* interpretation. The paper of 1870 also contains a substantial amount of material concerning the possibility of visualising non-Euclidean spaces. Helmholtz considers the by now familiar examples of beings living on the surfaces of spheres, and of transformed spatial relations in mirror worlds. Since these discussions do not add significantly to the philosophical thrust of the argumentation, I will summarise their significance briefly, and then turn to those critical passages at the end of the paper where Helmholtz addresses the interrelation of geometry and physics, and thereby first discusses the conventionalist position.

After illustrating the differences between Euclidean and non-Euclidean geometries by considering the concepts of straight and parallel lines when applied to convex and concave surfaces, Helmholtz demonstrates the compossibility of pseudo-spherical and Euclidean geometries by considering the "worlds" that present themselves in convex and concave mirrors. Since a mapping is possible between the two worlds, and since we are able to intuit the structure of both of these worlds with only slight effort, these examples show not only that there is no *logical* inconsistency involved in the notion of a non-Euclidean geometry, but also that it does not conflict with our pure intuition of space. The empty receptacle of our spatial intuition imposes no particular geometry on physical appearances. Although Helmholtz has now admitted pseudo-spherical geometries as equally admissible on logical and intuitional grounds, he can do so without much change to his earlier conception. For, on my reading at least, he had long regarded the properties of spatial intuition as restricted to dimensionality and continuity.

Because neither of the two competing geometries is to be preferred on *a priori* grounds, the central role accorded to rigid bodies in the 1868 papers is apparently given further support. There can be no talk of magnitudes without measurement, and measurement presupposes rigid bodies. However, as Helmholtz admits, the question of what counts as a rigid body cannot be answered without the aid of supplementary mechanical principles. It is here that he first considers the conventionalist option. He concedes that it is perfectly possible to consider "the space in which we live" as having the metrical relations that we find in a convex mirror, just as we could consider a limited portion of it to have a pseudo-spherical metric. Doing so would involve changing our physical laws entirely, for "even the proposition that every moving point not subject to a force moves in a straight line with unchanging speed does not apply in the world of the convex mirror."[28] Thus, Helmholtz concludes, geometrical axioms "do not speak about properties of space alone, but also about the mechanical behaviour of our most rigid bodies when they are in motion."[29]

28 (Helmholtz 1870), p. 29. See the section on "Über den Ursprung und Sinn der geometrischen Sätze" below.

29 (Helmholtz 1870), p. 30. This paragraph appears to have been inserted as an afterthought, as it does not so much advance the argument as abruptly introduce a possible objection. Furthermore the first passage sentence of the next paragraph "Die geometrischen Axiome sprechen nicht ..." does not follow logically from the conclusion of the paragraph at hand, whereas it fits quite well with the

Having shown, in his opinion, that geometry is necessarily a part of physical science (or as I have suggested, of phoronomy), Helmholtz quickly disposes of a possible Kantian objection, which would involve *postulating* the priority of Euclidean measurement standards. As in the conventionalist case, Helmholtz concedes that this epistemological stance cannot be rejected on the grounds of logical inconsistency; however, he claims that it is not consistent with Kant's own geometrical doctrines. For to accept this as a postulate would be to render geometrical propositions analytically true, whereas Kant held that they were synthetic *a priori.* At first glance, this may appear to be a trivial objection. For both Kantian and conventionalist opponents to Helmholtz's empiricist views should surely respond that the issue was not so much the synthetic character of the axioms (which could, after all, still be argued to be present in the assumptions of continuity and infinite extension) as their independence from experience. On their view, it is Helmholtz who has blundered into a vicious circle: measurement requires rigidity, rigidity is a dynamical concept, but the question of which dynamical laws hold cannot be separated from that of which geometry is valid. Thus there is no sense in which geometrical laws can be *independent* empirical truths, that is to say truths independent of other physical laws.

Whether or not Helmholtz's response to the "strict Kantian" is tendentious or not depends first, on the significance one attaches to Kant's claim that geometrical truths are synthetic, and, second, on what Helmholtz means by saying that his strict Kantian would make them analytic by postulating the validity of Euclidean geometry. Helmholtz, as we saw, regards analytic geometry as analytic in Kant's sense, for it expresses connections between pure magnitudinal *concepts.* I take it that his point is the following: by investigating the conditions on measurement by analytic means, we open the possibility of describing any possible manifold, and any possible measurement operation within such a manifold in analytic terms. That is to say, *if* there is a manifold that permits a given kind of measurement operation, then it is analytically true that the metric that will be ascribed to it under that operation will be thus and such. The Kantian who insists that the measurement operation must be the one he has chosen, has thus stipulated that the world in which he lives shall be described as if such measurements were possible, even if it turns out that every available apparatus behaves differently.

---

end of the preceding one "… ohne mechanische Betrachtungen hinzuzunehmen."

Helmholtz had, we may recall, criticised Graßmann's interpretation of his fourth colour-axiom on just this ground. On his reading of Graßmann, the latter had assumed that the results of his measurement-operation would conform both to the additive axioms, and to the circular form of his colour-wheel. Helmholtz had objected that, although one could indeed stipulate that such an agreement would hold, it was *at best* an arbitrary decision. There was no guarantee that the consequences of this decision would mesh with our experience in an intuitively coherent manner, for it might force us to call a red and a blue colour sensation equally "bright," even though we experienced them differently. In the case of geometry proper, it might result that the Euclidean metric that we postulated would define lines and lengths that were not compatible with the requirements of physics.

Helmholtz is correct in thinking that *Kant* would have rejected this conception of geometric truth. It makes geometry apodictic in the sense that its propositions are ordered by necessary logical relations; however, according to Kant, such an analytic stipulation has no claim to objective meaning, for there is no guarantee that anything actual corresponds to its concepts. The synthetic *a priori* truths of geometry, and of mathematics generally, are supposed to be binding on us not because they express a merely logical necessity, but because the relations they express are in some sense verified in intuition. When we construct their concepts, their truth follows necessarily, although not analytically. Kant did not consider the possibility envisaged by Helmholtz, for he assumed that the extensive structure of three-dimensional space entailed the truth of Euclidean geometry in some (non-logical) sense. That is why geometry can provide the base-level system of magnitudes for phoronomy. Such constitutive, mathematical truths, are distinguished from regulative propositions in their being *strictly* true, which is the same as saying that they are valid for *any possible* experiential data. He did of course recognise that the regulative requirements placed on natural science could be disappointed, but this failure could not derive from the inadequacy of geometry. Helmholtz, thanks to Graßmann, can envisage just this possibility, because he can separate the metrical properties that derive from additive operations with spatial magnitudes from the purely topological characteristics of an *n*-dimensional, extensive magnitude.

Thus we may characterise the shift in Helmholtz's arguments from the 1868 papers to the 1870 paper as follows. In the 1868 papers, the axioms that demonstrated the necessary truth of geometry were divided into regulative and constitutive groups. Because Helmholtz thought that these axioms singled out a single possible geometry, that geometry retained its transcendental status, but it was no longer constitutively required in the sense it was for Kant: the regulative demand that natural laws be maximally general defines the properties that any adequate empirical system of measurement must implement. And the algebraic analysis of these properties shows that *if* there is such an adequate system of measurement, then the system of magnitudes that they characterise will correspond to the axioms of Euclidean geometry. Once pseudo-spherical geometries are seen to conform to the regulative and constitutive demands equally well, the situation changes. Neither Euclidean nor pseudo-spherical geometry is singled out, but it remains the case that *one or the other* is regulatively required.

Helmholtz then identifies two options: we can suppose that the choice is conventional, and define "rigid body" in such a way that the one geometry is now necessarily mandated, but then we will end up with a geometrical theory whose truths are analytic; or, we can make the choice on inductive grounds. Helmholtz's suggestion is, in effect, that we have *already* made the latter choice, because the bodies that surround us conform more closely to the axioms of the one geometrical system than they do to the other. Thus Helmholtz forces the Kantian to choose between two equally unattractive alternatives: either the axioms of geometry are laid down independently of possible empirical intuitions, in which case they are apodictic, but analytic; or geometry is a species of phoronomy (a pure science of the motion of bodies). If one takes the latter option, one must admit that geometry is empirical. In the 1870 papers, this dilemma is already clearly articulated, however it is not phrased in the form of a refutation of Kant until the last paper of 1878, with which I will conclude this chapter. In the Conclusion, I will reappraise the significance of these papers from the point of view of the philosophy of physics, in particular the *Conservation* and the theory of electrodynamics.

(d) "Über den Ursprung und Sinn der geometrischen Sätze" (1878)

As I outlined briefly in my introductory remarks to this chapter, this last paper of Helmholtz's on geometry was first published in English in *Mind* as a response to the objections of J.P.N. Land, a professor of philosophy at the University of Leyden.[30] His critique of Helmholtz's 1870 paper, which appeared in *Mind* under the title "Kant's Space and Modern Mathematics,"[31] was a not unsympathetic attempt to evaluate the significance of cutting-edge developments for the philosophy of mathematics. As the title indicates, Land was a confirmed Kantian, and it is therefore his Kantian reaction to Helmholtz that determined in large part the anti-Kantian thrust of Helmholtz's reply. This rebuttal was first published in its original German version only in the *Wissenschaftliche Abhandlungen*, while an abbreviated version of it appeared as the second appendix to Helmholtz's *Die Tatsachen in der Wahrnehmung*, the philosophical manifesto that he delivered as a public lecture in the same year. I will discuss only the longer version of this paper in the following. Since the paper appeared eight years after the preceding one, it is important to keep its arguments just as distinct from them as we kept the position of the 1870 paper distinct from the ones from 1868. In order to correctly situate Helmholtz's arguments, these will be related both to those of Land's, as well as being contrasted to the earlier geometry papers and to the work from the 1850's that has been discussed in the earlier chapters. Before summarising briefly the content of Land's paper, I must, however, address two slight modifications to the text of the 1870 paper which Helmholtz appears to have made on the occasion of its 1876 publication in *Mind*.[32] These changes are important for our understanding of the concept that was first explicitly defined in the 1878 paper we shall be considering, namely that of a *physical geometry*—a concept whose importance for the development of the theory of relativity is well known.[33] Both of these two additions are made to the concluding passages of "Über den Ursprung und Bedeutung." The first concerns the relation between geometrical and physical laws, and the second articulates the dilemma that I described at the conclusion of the last section, which had not been clearly spelled out in the first, German version of the paper.

In the original, 1870 version of "Über den Ursprung und Bedeutung," Helmholtz allows that one could conventionally choose among the two admissible (Euclidean or pseudo-spherical) geometries if one found this to be expedient. He then observes that this choice might entail changes in properly physical propositions, for if we impose a non-Eucli-

30 (Helmholtz 1878b)

31 (Land 1877)

32 (Helmholtz 1876)

33 *Cf.* (Friedman 2002).

dean metric on our (Newtonian) world it would follow that the speed of an inertially moving body would change with its position. As I mentioned in the Introduction, this passage has been taken to show that Helmholtz thought that by appealing to physical laws, one could single out one geometry as empirically valid.[34] In doing so, he would have overlooked the fact that such physical laws were themselves conventionally adjustable, and thus he would also have failed in his avowed purpose, which was to shore up his earlier claim that geometrical propositions were empirical statements concerning the behaviour of rigid bodies. But this reading does not adequately consider Helmholtz's reasons for appealing to supplementary physical laws. Furthermore, it overlooks a fundamental shift in Helmholtz's understanding of what counts as a spatial magnitude, a shift occasioned by the conventionalist possibility he identifies here. Thus it is important to keep the arguments before and after this break distinct. Helmholtz had, in 1870, only partially comprehended the significance of the conventionalist objection. And so he still argues on the assumption that rigid body displacements are sufficient to circumscribe the concept of a spatial magnitude. According to the 1878 paper, however, the business of geometry is to define what Helmholtz calls "physically equivalent" magnitudes, which are magnitudes in which physical processes take place in the same period of time (for instance equal distances along inertial paths). According to such a definition, geometry is "empirical" in a quite different sense from that which Helmholtz originally envisaged.

That this shift has been overlooked is, however, understandable, because the chain of thought that leads to the new definition is incomplete in the 1870 paper. There, Helmholtz merely observes that we *could* choose to reconfigure our geometrical and physical laws if "we found this useful to some purpose," and that this would be "perfectly consistent." On the next page, he emphasises that it is only when the axioms of geometry are combined with mechanical principles that "such a system of propositions acquires a real content." Thus it may sound as though the arbitrariness of our choice can be eliminated by invoking the empirical content of higher-level laws. But there is no need to assume that he thinks that, for, as I have just explained, his main objection (which is stated immediately following this passage) is that geometry would thereby be rendered analytically true. Moreover, this is not the reading that Helmholtz later gave of this appeal, however he may have meant it in the original

34 See for instance (Torretti 1978), p. 169, (Carrier 1994), p. 282, (Schiemann 1997), p. 233.

text. In the 1876 English version, this critical paragraph is supplemented by two sentences, which are given here in italics:

> But if to the geometrical axioms we add propositions relating to the mechanical properties of natural bodies, were it only the axiom of inertia or the single proposition that the mechanical and physical properties of bodies and their mutual reactions are, other circumstances remaining the same, independent of place, such a system of propositions has a real import which can be confirmed or refuted by experience, but just for the same reason can also be got by experience. *The mechanical axiom just cited is in fact of the utmost importance for the whole system of our mechanical and physical conceptions. That rigid solids, as we call them, which are really nothing else than elastic solids of great resistance, retain the same form in every part of space if no external force affects them, is a single case falling under the general principle.*[35]

We have encountered this “general principle” that “the physical properties of bodies and their mutual reactions are independent of place” elsewhere in our reading of Helmholtz. It is another version of the claim that the physical descriptions of systems should not involve relations to absolute space. In the case at hand, the suggestion is that by choosing a non-Euclidean metric, we will have to admit laws of motion which make the physical properties of systems depend on location, and this speaks against such a geometry. Helmholtz is therefore arguing not that the laws of mechanics decide empirically between competing geometries, but rather that we have systematic reasons for preferring theories in which laws are positionally independent. In positing a non-Euclidean metric in a Newtonian-Euclidean world, a conventionalist would force us to make the same sort of error as electrodynamic theorists who admit relations to absolute space. We would end up with a theory that was mathematically possible, but nevertheless transcendentally inadmissible. We have, in other words, systematic epistemological reasons for preferring a metric that supports position-independent laws.[36]

The epistemological justification for the appeal to physical laws is, however, only half of the story. A natural interpretation of these key pas-

35 (Helmholtz 1876), p. 320, my emphasis. The unitalicised passage is in the German original as well, (Helmholtz 1870), p. 30.

36 The reader may object that I am ascribing a position to Helmholtz that is almost indistinguishable from Poincaré’s. But that is not the case. For whereas Poincaré claims that we will always employ Euclidean geometry because it is more convenient, Helmholtz thinks that it is epistemologically unjustifiable to introduce positionally dependent laws. It is not merely that these are inconvenient, but rather that they introduce what we can only comprehend as transcendent causes, and thereby violate the requirement of the comprehensibility of nature.

sages is that Helmholtz is arguing that geometry is empirical because we must opt *either* for a pseudo-spherical *or* for Euclidean geometry once we have accepted the law of inertia in its standard, position-independent form. Thus his aim is to show that the two options represent distinct and contingent possibilities among which we must choose on the basis of empirical data. But this ignores the full context in which these remarks appear, namely the dilemma that I sketched above. Geometry, in order to be physically applicable, must codify the properties of those empirical systems used to define spatial magnitudes. Kantians must therefore choose between the following alternatives: (1) the set of rigid bodies is *defined* as those whose coincidence relations satisfy the axioms of Euclidean geometry, from which it may follow that no (even approximately) rigid bodies exist; or (2) it is not. If they choose option (1), the resulting geometry is analytic, thus it does not describe synthetic *a priori* characteristics of space.[37] Thus they must opt for (2), and this will mean giving an account of what is to count as a rigid body that does *not* rest on purely conventional grounds.

Now, it could be that having chosen (2), we could still show that Euclidean geometry is the only possible science of spatial measurement adequate to our regulative requirements. That is just what Helmholtz thought he had done in the 1868 papers: there it looked as though these requirements could be met *only* by a set of bodies whose coincidence relations satisfied the Euclidean axioms. Thus the independent criterion of what was to count as a rigid body would have been given by appeal to regulative demands placed on spatial measurement. This would indeed have been a synthetic *a priori* demonstration of the necessity of Euclidean geometry, even though the necessity would have been regulative, and not constitutive. But, it turns out, this is not possible: the regulative demands can also be fulfilled by bodies satisfying pseudo-spherical relations. This means that both sets of geometrical axioms are equally admissible, *even if* they are interpreted as statements concerning the coincidence relations of empirical bodies. Only their conjunction with (at least) the law of inertia results in a system that has "real import" (*einen wirklichen Inhalt*).

As we shall see in a moment, Helmholtz eventually concludes that there is no science of spatial magnitudes *per se.* For such a science, in

37 Note that this procedure is quite close to that followed by Helmholtz himself in the Königsberger manuscript. It is, furthermore, the sort of "conventionalist" approach that Helmholtz criticised in Graßmann.

order that it might play the role cast for it within physics, would have to provide the basic stock of relations that are invoked in our formulation of kinematic and dynamic principles. And these must be, as always, magnitudes that are metrologically realised, in the sense defined in the 1868 papers. By adding in the requirement that basic spatial magnitudes describe kinematic properties, Helmholtz effectively eliminates geometry as an autonomous science. In Kantian terms, we may say that the definition of a spatial magnitude has been rendered wholly phoronomic: equal spatial magnitudes are ones traversed by a body in inertial motion in the same time interval. Nevertheless, Helmholtz continues to oppose himself to Kant's theory, in which the additive properties of the basic kinematic magnitudes were to be constructed in pure intuition. For Helmholtz, such constructions are necessarily indeterminate, for the inherent properties of space could not form an object of a possible experience. And thus the requirement that there exist bodies which satisfy the basic measurement axioms (on the model of Graßmann's additivity axioms) remains in force.

From the point of view of later conventionalists, Helmholtz's reasoning here is flawed because he thought that there was a fact of the matter as to whether inertially moved bodies do, or do not, traverse equal distances in equal times.[38] They maintained, on the contrary, that by suitably readjusting mechanical laws, we can accommodate the observed phenomena to various geometries. But I do not think that Helmholtz would have been much impressed by this response, because such adjusted mechanical laws suppose, e. g. that the law of inertia is positionally dependent, or indeed that there are positionally dependent universal forces.[39] Helmholtz has already immunised himself against this strategy in the passage cited above, for he insists that all laws must be positionally independent. And this is not an *ad hoc* requirement on Helmholtz's part. He had insisted that positional independence "is … of the utmost importance

38 Comparing Helmholtz's position to that of Carnap and Reichenbach, Martin Carrier observes that, "[t]he gist of the conventionalist's reply to Helmholtz thus consists in the contention that it is not a fact of the matter that inertially moved bodies traverse equal distances in equal times. Rather, as conventionalists insist, the law of inertia can readily be reformulated to the effect that the velocity of bodies depends in a specified way on their location." (Carrier 1994), p. 283.

39 That is to say, in order to reinterpret the events of what Helmholtz calls "our" world (a world in which Newtonian Mechanics holds in Euclidean space) in terms of a non-standard metric, we have at some point to introduce position dependent laws in order to compensate for the divergence between the two metrics.

for the whole system of our mechanical and physical conceptions" in the *Conservation* as well. By contrast, Carnap's later demonstration that we can adjust our physical laws to conform to an arbitrary geometry led to the result that energy would not be conserved in the new metric. Carnap could only avoid this conclusion by introducing a conventional definition of energy that explicitly depended on position.[40]

Thus Helmholtz is clear on the fact that one could interpret "our" Newtonian-Euclidean universe as having a non-Euclidean metric, in the sense that one could *stipulate* such a metric. But this would force us to "correct" the deviant behaviours of our measuring instruments by means of position-dependent physical laws. The logical possibility of conventionally adopting the one geometry or the other does not, on Helmholtz's view, change the fact that the significance of geometric propositions depends on their physical role in kinematics, and not in their representing a science of space *per se.* But it does show that the proper account cannot be the one that Helmholtz insisted upon in the 1868 papers. A new interpretation, only partially realised in the 1870 paper, is concretised in the new concluding paragraph of its 1876 translation:

> To sum up, the final outcome of the whole inquiry can be thus expressed:—
>
> (1.) The axioms of geometry, taken by themselves out of all connection with mechanical propositions, represent no relations of real things. When thus isolated, if we regard them with Kant as forms of intuition transcendentally given, they constitute a form into which any empirical content whatever will fit and which therefore does not in any way limit or determine beforehand the nature of the content. This is true, however, not only of Euclid's axioms, but also of the axioms of spherical and pseudo-spherical geometry.
>
> (2.) As soon as certain principles of mechanics are conjoined with the axioms of geometry we obtain a system of propositions which has real import, and which can be verified or overturned by empirical observations, as from experience it can be inferred. If such a system were to be taken as a transcendental form of intuition and thought, there must be assumed a pre-established harmony between form and reality.[41]

Whether or not he had actually drawn this conclusion in the original version of the 1870 paper, Helmholtz was now aware that the argument of the 1868 papers could no longer be sustained in its original form. The mere possibility of comparing distances by means of consistently coincident bodies does not yet yield a system of magnitudes that has a physical significance (a "real import" or *wirklichen Inhalt*). And, from a strictly

40 *Cf.* (Carnap 1922), pp. 52–53.
41 (Helmholtz 1876), p. 321.

formal point of view, any set of axioms that ascribe to space a constant curvature determines a "form of intuition" that can accommodate arbitrary empirical intuitions, thus a structure of experience agreeing to Kant's notion of spatial intuition. But by appealing to a strong form of positional independence, we can still place demands on geometry that are strong enough to force our hand. The axioms will then circumscribe the properties of those bodies used for measurement. These in turn determine spatial magnitudes adequate to the description of all physical phenomena by means of positionally independent, general laws of nature. Such magnitudes are defined at the beginning of the second section of the 1878 paper as "physically equivalent" magnitudes. In Helmholtz's Newtonian-Euclidean world, the preferred axioms are those of Euclidean geometry. But we can also imagine a world in which pseudo-spherical axioms would yield a more general system of physical laws.

Thus by the end of the 1876 version of this paper, Helmholtz has shifted his position substantially. Pure geometry describes a set of relations that hold of any empirical content that appears in space. Thus there could be multiple pure geometries, each of which is consistent with space containing arbitrary appearances. Whereas Kant would have said that the pure geometric (for him, Euclidean) relations *must* hold of any spatial phenomenon, we must now say that they *can* hold of any spatial phenomenon. There is no immediate impediment to imposing either a Euclidean or a non-Euclidean metric. If, on the other hand, these relations are connected to positionally independent mechanical axioms, most notably the law of inertia, this is no longer the case. For the choice of the one geometry, once the lines it defines are interpreted as inertial paths, effectively excludes the other.[42] Just as in the Königsberger manuscript, where Helmholtz drew a distinction between the non-motive and motive properties of space, here he distinguishes between spatial relations considered in and of themselves, and the same relations considered as basic kinematic magnitudes. The lines and distances of each geometry are now identical with the lines and distances that Kant introduces in the Phoronomy of the *Metaphysical Foundations.*

This change in Helmholtz's views is concretised at the outset of his 1878 reply to Land, where he first introduces the notion of "physically equivalent magnitudes": two magnitudes are physically equivalent when the same processes may "exist and unfold" (*bestehen und ablaufen*) within

42 Once again, if the law of inertia is similarly adjusted, other laws, such as energy conservation will be violated, or be rendered position-dependent.

them under the same conditions and in equal periods of time. The "most commonly used process for determining physically equivalent magnitudes," he goes on, "is the transport of rigid bodies, such as compasses and rulers, from one location to another."[43] The equivalence relations that define sets of physically equivalent magnitudes are distinct of those "which derive from transcendental intuition" in that only the former are determinable "with physical instruments." Although he does not explain further what exactly is meant by "the same physical processes," Helmholtz's meaning is clear enough in the light of the passage from the 1876 translation quoted above. There, he claims that a system of geometrical axioms gains an empirical meaning when it is supplemented by mechanical principles, such as the law of inertia, or the still more general principle that mechanical laws and physical properties are independent of place, in other words what I have been calling positional independence. Borrowing an example from Carnap,[44] we may imagine determining such physically equivalent magnitudes by means of a spring-loaded ball-launcher, such as those used in pinball machines. The device has a mark on it to tell us when the spring is set to "the same" tension. We then fire it at various points in space, and mark the point the ball has reached after a given time-interval, which could be defined as a set number of oscillations of the plunger after the ball has been fired. Such a pinball launcher would then be a transportable device for operationally defining physically equivalent magnitudes.

In his comments on this passage in his edition of Helmholtz's *Epistemological Writings*, Schlick suggests that Helmholtz's definition is question-begging, because it depends on a time-standard that can only be determined with reference to established distance-standards.[45] Thus Schlick would object that the proponent of a pseudo-spherical metric could reinterpret our operational, ball-launcher determination of physically equivalent magnitudes as follows. "You claim that the ball-launcher's behaviour at each point in space satisfies the concept of 'same process under same circumstances.' For you, the path of the ball during n oscillations of the plunger represents a magnitude that is physically equivalent to all other paths so determined. But I claim that the physically equivalent

43 (Helmholtz 1878c), p. 648.

44 (Carnap 1922), p. 52.

45 (Helmholtz 1977), p. 183. The original German version (Helmholtz 1921) has been reprinted as (Helmholtz 1998), in which the passage in question is on pp. 225–226. The criticism is reiterated in (Hatfield 1990), p. 222.

magnitudes are those of a pseudo-spherical geometry. On my view, there is a global force acting in space, and the underlying geometry is pseudo-spherical. Your 'inertial' paths are actually the paths of forced motions. And even though the number of oscillations observed is the same for all your measurements, the speed of the plunger is different at different points in space: once you correct according to my metric, you will see that your physically equivalent magnitudes are in fact physically diverse." One could then argue that the positions of the Euclidean and non-Euclidean interpreters of the phenomena are symmetric, for if the non-Euclidean proponent could come up with an apparatus that generated *his* set of physically equivalent magnitudes in an analogous manner, his Euclidean opponent would defend himself in the same way.

But, as I suggested above, Helmholtz has a response to this objection at hand, and it is indeed the same response that motivated him throughout the entire sequence of his writings on geometry, from the Königsberger manuscript on down. Whatever the exact content of our fundamental laws, they must be positionally independent. Since Helmholtz assumes that Newtonian mechanics in a Euclidean universe offers at least an empirically adequate characterisation of our experience, he also assumes that any alternative geometry would—in this (for Helmholtz, Newtonian-Euclidean) world—result in the appearance of positionally dependent laws at some point in the system. The introduction of global force-fields, or of positionally dependent changes in the energetic state of systems would simply be one more case of transcendent causation. His demand remains the same: strict positional independence must be coupled to the requirement of empirical determinacy. Thus we must use empirically realised standards of length that do not result in positionally dependent laws. Helmholtz is in other words fundamentally committed to the uniformity of space—a commitment which is also reflected in his restriction of Riemann's analysis to spaces of constant curvature. Any theory which involves spatial dependencies is to be rejected on *a priori* grounds.

Having secured a definition of physically equivalent magnitudes which defines the group of inertial motions, Helmholtz moves in the second section of the paper to his direct refutation of his Kantian opponent, Land. He argues that the group of physically equivalent magnitudes *need not* coincide with the group of equivalent magnitudes that derive from transcendental intuition—but the Kantian must assume that they do coincide. Helmholtz defines two possible species of geometry: a pure intuitive one, which determines relations of equality between distinct spatial magnitudes by means of direct intuition, and a physical geometry, in which the relations of equality are those which accord with the definition of physical equivalence. He contends that these sciences can be pursued independently, and that their results need not agree. The geometry that results from judgments based on immediate intuition has no necessary connection to the physical geometry, precisely because it refers neither to matter nor to time. Such a geometry, as Helmholtz had already explained at the conclusion of the 1876 translation to which Land had responded,[46] "represent[s] no relations of real things." But no Kantian will be prepared to concede that pure geometry is scientifically irrelevant, and so the latter must explain how and why this geometry can be said to play a role in physics at all. Since Helmholtz's competing physical geometry is *defined* in such a way as to satisfy the demands placed by physics on its fundamental magnitudes, the Kantian must show that the two geometries will agree, indeed that they must agree. Helmholtz's claim is that there can be no demonstration of this necessary agreement—if the two geometries do agree to one another, this can only be the result of a pre-established harmony.

To make his case, Helmholtz must therefore prove the absence of a necessary connection between pure intuitive and physical geometries. The argument he offers is direct. He assumes that we have found a method (say our ball-launcher method) to determine physically equivalent magnitudes. Using this realised definition, we can then define the notion of distance, and by means of the latter, the notion of a straight line. Given these physically realisable definitions of straight lines and distances, we can construct spatial figures and verify whether or not they satisfy the relations codified in a given geometry. But since even the Kantian must admit that the results of the requisite measurements are contingent, it follows that one cannot say in advance which geometry will be verified. Indeed, there is no guarantee that any will be verified.

Even though his argument for the empirical character of geometry is not as direct as it was in the 1868 papers, the fundamental epistemological point is not much different, indeed it is more obviously connected to concerns raised in the *Conservation* project. The fundamental magnitudes of the physical geometry are nothing other than the magnitudes postulated in pure kinematics. From an epistemological point of view, such a science could resemble classical geometry, that is a science of spatial magnitudes that is independent of physical bodies, only to the extent that the magnitudes involved were found to be physically equivalent for all proc-

46 See the first paragraph of the quotation on p. 190.

esses found in nature. In general, if it were the case that the science of physical geometry was found to agree with pure intuitive geometry, this could only be a happy coincidence. Furthermore, pure geometry could have no practical utility, for physics would always appeal to the magnitudes of physical geometry when formulating its laws. Whatever one imagines the science of these pure magnitudes to be, still it remains the case that they must be realised if they are to be applicable to physical reality. In the reply to Clausius from 1854 and at the conclusion of the 1876 translation, Helmholtz phrases the point identically: the magnitudes involved must be "relations among real things." A space with no bodies in it cannot have a metric, for there are no comparisons to be made. Furthermore to imagine that it did have one adds no physical value, for that metric could never be an object of possible experience. Helmholtz remains true to his original intuition, which was that geometry is unavoidably concerned with fundamental *motive* determinations of space.

In the concluding chapter, we shall explore the link between the geometry papers and the *Conservation* programme in greater detail. But the central connection is already evident in the light of the above discussion. In both cases, Helmholtz pairs a regulative demand (nature must be described in terms of general laws) with a constitutive demand (space is undifferentiated, therefore such laws may not be positionally dependent), in order to exclude magnitudes of a specific kind from physical theory. In the energy dispute, the magnitudes thus excluded are non-central forces. The exclusion rested on the claim that the energetic state of a system is the same whenever it is in the same internal configuration—any dependence on absolute position is disallowed.[47] Thus it followed (fallaciously) that the forces acting within the system had to be centred on those same empirical points whose relative positions determined the internal state. In the geometry papers, the magnitudes excluded are "pure intuitive" magnitudes, whether these are taken to result from conventional stipulations or from an unmediated intuition of the structure of space. In order to qualify as empirically meaningful, spatial magnitudes must be determined by material points, and these magnitudes must permit the description of motions by means of completely general laws. By means of this second constraint, Helmholtz can argue that we should accord empirical significance only to those systems of geometry which do

47 More precisely: if the principle of energy conservation is to be empirically determinate, then it must be interpreted in this manner.

not violate positional independence, in the sense that they do not force positionally dependent mechanical principles on us.

The shared features of these arguments are thus, (1) the transcendental determination of possible magnitudes by "squeezing" them between regulative and constitutive demands, (2) the attempt to eliminate all reference to inherent properties of space beyond dimensionality and continuity. This resemblance is more than a mere analogy. The principle of the conservation of *vis viva*, if it is to be empirically determinate, requires that we have a means of comparing the states of a system at different places and points in time. It says that the energy state of the system is the same *whenever* it is in metrically identical states, from which it supposedly follows *a priori* that there can be no positionally dependent forces. Conversely, according to Helmholtz's criterion for selecting empirically meaningful geometries, any metrical convention that results in the assumption of positionally dependent laws is to be rejected. In both cases, the postulate that there can be no such forces does not mean that this is *logically* impossible. Our experience could be such that we had no choice but to assume them. But doing so one would be in essence to admit failure, in that we would have accepted a form of transcendent causality, for we could give no sufficient reason for the asymmetry in question. In both cases, Helmholtz is concerned with the threat posed by the concept of absolute or, one might say, "free" forces, to the project of completely determining nature. Newton took the existence of forces as a criterion for distinguishing apparent from absolute motion. The Kantian tradition to which Helmholtz belongs sees it the other way round: there is no absolute motion, and thus there cannot be absolute forces either. Nature must be, as I shall explain in conclusion, *empirically closed.*

7. Conclusion

According to one influential reading of Helmholtz's work on geometry, he argued for its empirical character because of his opposition to nativist theories of spatial perception. The papers on geometry then represent above all a contribution to sense-physiology and naturalist epistemology, and are only secondarily concerned with the role of geometry in physical science. This view, which has been most forcefully argued for by Hatfield,[1] has much to recommend it. For there can be no doubt that Helmholtz had an ongoing concern with establishing the validity of an empiricist epistemology over and against nativist theories, which assumed that the connections established among the various data of the senses were in large measure pre-programmed. The compossibility of Euclidean and non-Euclidean metrics provides strong evidence for the thesis that apparently fundamental aspects of spatial intuition are actually established through experience, and are not *a priori* properties of a matrix for ordering experiences. Moreover, Helmholtz himself emphasises the importance of his work in sense-physiology as a background to his work on geometry. Both the mathematics and the epistemological interpretation he gives of his formal demonstrations are, as I have sought to illustrate in Chapters 4 and 5, fundamentally determined by the work in physiological optics.

Nevertheless, this interpretative approach can lead one to overlook an entirely separate chain of reasoning that is at work in these writings—one which is, if I am not very much mistaken, the actual source of the *arguments* that Helmholtz offers. Here the central concern is with the role of geometry as a science of spatial *measurement.* Geometrical propositions are *empirical* not so much in the sense that they are *inductive* as that they are *material.* The criticism directed against Kant is that he conceives of space as a system of magnitudes that can never be objects of experience. Since, however, natural science deals with of the relations among real things, all of its propositions should concern material, that is to say, empirically given states of affairs.

1 (Hatfield 1990)

The strongest evidence in favour of the reading I am suggesting is offered by the transcendental arguments of the first two, technical papers. As I argued in Chapters 5 and 6, these papers present a transcendental proof of the validity of Euclidean geometry. But what is claimed is not, as in Kant, that Euclidean geometry is synthetically *a priori* true because it derives from constitutive requirements on possible experience. On the contrary, in his 1868 papers, Helmholtz argues from regulative requirements to prove that, if it is possible to measure all bodies in space, then space must have a Euclidean structure. The transcendental requirement results from highly restrictive demands issuing from the physical sciences: it must be possible to characterise relations among spatial appearances by means of mathematics, for otherwise we could not establish general physical laws. Furthermore, these demands are regulative: if they are not met, it follows only that such general mathematical descriptions of nature are impossible. Now, the quantitative comparison of different regions of space requires that there be bodies in space that can be displaced, and indeed partitioned. Such bodies must be material if they are to be used to measure the sizes of other bodies, and if they are to be used to describe the inertial paths of physical systems. But whatever is material is an object of perception, and the course of our perceptions is what Helmholtz calls a matter of fact. Thus the science of spatial measurement presupposes the existence of certain kinds of material facts. Conversely, in so far as we act on the assumption that descriptions of this sort are possible, we also tacitly assume that spatial measurement is possible, and accordingly that bodies satisfying the axioms of Euclidean geometry can exist. The axioms of Euclidean geometry are necessary, in the sense that if they cannot be satisfied by some class of objects, our scientific needs will be frustrated.

There are two essential premises involved in this argument: (1) whatever is material is contingent; (2) only material relations are observable. So long as one assumes that measurement is a form of observation, it follows from (2) that the instruments we utilise in carrying out spatial measurements are material bodies. And then, from (1), we can infer immediately that the existence of these bodies is a contingent fact. It should be emphasised that both of these propositions are deniable. One might, for instance, claim that spatial properties are properties of material objects that are not contingent, except in the degenerate sense that they have no bearer when no material objects are at hand. Alternatively, one might contend that certain non-material relations can be "observed," in the sense that they can be *intuited* without being *perceived*.

In fact, Helmholtz admits all along that *some* properties of material systems are not contingent, for he never abandons the idea that some spatial properties—at the very least dimensionality—represent formal conditions on external experience. Thus in the later papers, he does not so much sustain premise (1) as widen it: he denies that what counts as a change in the state of a body according to an orthodox, Euclidean theory of space, *must* be regarded as a change;[2] or, more simply, he contends that not all the axioms of Euclidean geometry refer to necessary properties of physical appearances. But he is quite clearly opposed to (2), namely the idea that there can be non-perceptual observation. The only position that is consistent with Helmholtz's take on (1) and (2), but which nonetheless contradicts his claim that geometry is empirical would then be the following: the measurement of spatial relations must rest on perceptual evidence; however, some of the relations that are *observed* are *formal* properties of material bodies; lastly, these formal properties include those relations among appearances that we codify in geometrical propositions. A neo-Kantian holding this line could then argue that, although measurement is accomplished by comparing *material* objects with one another, the lengths of the objects involved are formal, and not contingent material properties. Helmholtz himself appears to have taken this position in the Königsberger manuscript, where he appealed to our ability to intuit the identity of bodies undergoing displacement in order to explain rigidity. The length of a body would thereby be a necessary property of the latter, in that we would simply deny that it was "the same" body when the length changed. Measuring bodies that were indeed subject to change (truly physical, as opposed to ideal mathematical bodies) would therefore consist in comparing them to bodies whose properties did not change at all when they were displaced. Such unchanging bodies are rigid bodies. They represent a limiting case of physical bodies.

On closer examination, we can see that this interpretation differs only slightly from that advanced by Helmholtz in 1868. For the metrical relations among the rigid bodies he considers there are obviously not intended to be empirical in the sense that he holds that there actually *are* perfectly rigid bodies. On the contrary, the existence of such bodies is a normative demand placed on our system of spatial measurement by natural science. Helmholtz's claim is therefore not that we have inductive grounds for believing in the existence of such bodies, and thus inductive grounds

2 He does this when he acknowledges that one *could* insist on using the one or the other geometry.

for believing in the truth of geometry. His point is rather that *whatever it is* that satisfies the axioms of Euclidean geometry, it cannot be conceived as consisting of the relations among the parts of empty space. The relations satisfying geometrical propositions are empirical because they are indivorcibly connected to the motive characteristics of material bodies, and not primarily because these characteristics are inductive truths, even if that thesis may follow from the latter. Indeed, he emphasises this distinction on the first page of the 1868 "Über die tatsächlichen Grundlagen der Geometrie," where he points out that his investigation is "quite independent of the further question concerning the *origin of our knowledge* concerning these propositions with factual meaning."[3]

Thus I would submit that there is no sense in which these first two papers are concerned with establishing the inductive character of geometrical axioms, even though they are very much concerned with establishing their material character. Since Helmholtz does, in the later two papers, proceed to argue that geometry is empirical in both these senses, this peculiarity of the first two papers has generally been overlooked.[4] But this shift in his position does not free us from the responsibility of explaining his motives for arguing as he does here. We cannot simply ignore this aspect of his thinking as if it were a mere anomaly. As I have suggested in this study, we have ample evidence for the claim that Helmholtz's concern with the *material content* of geometry was independently motivated. Furthermore, Helmholtz would not be the first to regard the kinematic core of mechanics to be an *a priori* science of motion. Thus it is not so surprising that the dichotomy between empirical and formal truths does not always coincide with the distinction between inductive and *a priori* truths in Kant's system, since certain empirical (in the sense of material) truths can be *a priori*. And despite the emphasis he comes to lay on the inductive background that may have given rise to the one or the other system of geometry, Helmholtz's essential move consists in demonstrating that geometrical truths must be understood as statements concerning the motive characteristics of rigid bodies, or, more generally, as statements concerning the properties of space that are required for it to "comprehend all possible *changes of matter*."[5]

To take a somewhat anachronistic view, we might say that Helmholtz is opposed to the idea that there is a pure part of natural scientific knowl-

3 (Helmholtz 1868a), p. 610. My emphasis.
4 A notable exception is (Dubucs 1998), p. 113 ff.
5 Helmholtz in (Königsberger 1903), p. 134. My emphasis.

edge which is *not* empirical (in the sense of material), although he is prepared, at least up until 1868, to admit a part which is both empirical and pure. This philosophical position is not motivated as much by empiricism in the sense of a commitment to inductive methods, as it is by a commitment to a form of verificationism, albeit one that is strongly Kantian in flavour. Helmholtz believes that mathematics is a science of magnitudes. Whatever mathematicians may imagine when working with a pencil and paper, their calculi can only be applied to physical reality when we have adequately characterised a domain of magnitudes for which they hold. We must be careful that the power of the mathematics not lead us to posit relations which do not have a definite empirical correlate. But this is just what Kant and his followers have done, according to Helmholtz. They argue that the truths of geometry and arithmetic depend on intuitions which, even if they are supposed to be mere forms of empirical experience, are nonetheless independent sources of evidence. As a consequence, this way of thinking can lead us to confuse purely mathematical properties with physically determinate ones. In the worst case, it can lead us to introduce relations into physics which are not the contents of a possible experience.

Thus on my reading, the logic of Helmholtz's 1868 papers is directly related to that of the Königsberger manuscript analysed in Chapter 5, as well as to the arguments from empirical determinacy advanced in the *Conservation.* But even if one denies the latter connection, it remains the case that one cannot explain the motivation of these papers by referring to Helmholtz's commitments to an inductive programme. Furthermore, as I will consider briefly in the following, there is very little in the two papers following (those of 1870 and 1878) which contradicts my reading. Although Helmholtz does suggest that beings living in different universes would, on inductive grounds, develop alternative systems of geometry, that is not the principal argument that he directs against Land and others. Indeed, he concludes the last paper with a variant of the indeterminacy argument that he had directed against Clausius: even if it were true that we could intuit the magnitudinal relations among material things, this knowledge would be of no use to us, since we would always prefer the magnitudinal relations that we had set up in empirical geometry. Non-empirical, purely mathematical knowledge, may well be possible; however, it can never be *actual,* to borrow Kant's terminology, because the magnitudes it treats of are not properties of real things.

Kant, we may recall, tries to steer a middle course. While he holds that mathematical knowledge is not analytic, in that it does not derive from purely logical reasoning, he also contends that it is not empirical. Mathematical concepts are provided with corresponding intuitions by means of *a priori* constructions. These schemata give mathematical propositions a real content which is nevertheless not empirical. This theory seems, on the face of it, to answer the question posed in the first *Critique* and the *Prologemena*, namely how synthetic *a priori* knowledge is possible. And thus it ought also answer the corresponding question in the philosophy of science: Why is pure mathematics applicable to empirical reality? This question, after the advent of Newtonian mechanics, is even more pressing than it was in its classical form. For although the relation of ideal mathematical objects to their material counterparts is a locus classicus of philosophical debate from Plato onwards, it takes a far more radical form in the philosophy of early modern physics. The specific question becomes: Why are geometrical propositions simultaneously kinematic propositions? That is to say, why are the paths of inertially moving bodies straight lines? Why do forces add according to the parallelogram law, that is to say "geometrically"?

I argued in Chapter 2 that Newton has a two-fold answer to these questions. The first, empiricist answer is that there is no necessity at all at work here. Both the law of inertia and the second law of motion (which yields the parallelogram law as a corollary) are empirical propositions, whose justification is inductive only. But this conservative position presumes a second, more generous one, namely that space is absolute, and that geometry is a science of its properties. So while he holds that it is an empirical fact that inertially moving bodies follow straight-line paths, Newton is not troubled by doubts concerning the determinate *meaning* of this proposition. There may be empirical difficulties in ascertaining whether an isolated system is moving inertially with respect to absolute space or not—indeed these difficulties may be, in principle, insurmountable—but it remains the case that, from a bird's eye view, there is a fact of the matter here. It is only by accident that we can never verify such a proposition.

Kant disagrees with Newton on both counts. He holds that the notion of an absolute space is incoherent. And he believes that he can offer *a priori* proofs of both of the law of inertia and of the parallelogram law of forces. The first of these claims follows directly from the doctrine that space is a form of outer experience. It is true that we can "imagine space without there being any objects to encounter in it";[6] however,

6 *KrV* A24/B39.

this empty space is completely indeterminate, therefore it cannot be an object of a possible experience. For outer experience presupposes not only a form, but also the matter that fills it. Then, and only then, can we speak of determinate magnitudes and determinate relations in space. Kant then utilises this very empirical indeterminacy to prove the apodictic character both of the law of inertia and of the kinematic and dynamic parallelogram laws, in all cases fallaciously. According to Kant, we need no law of inertia because there is no empirical difference between straight-line motion and motionlessness. To "construct" the one is to construct the other, since in both cases the only experientially determinate relation is a change in the length of the straight line connecting two points. The parallelogram laws of motion and acceleration composition, and thus that of force composition, all derive from similar demands. For these concepts can be constructed only as changes in the relative positions of empirical points, which Kant then uses to derive the necessity of decomposing momentum change geometrically, into the sum of the actions of central forces. In other words, although Kant does not believe that there can be an empirical concept of absolute space, he is convinced that the application of geometry to basic physical processes is apodictically required. Indeed, we can squeeze physical principles out of this demand—principles that Newton thought were empirical. But although he has justifiably posed the question concerning the role of geometry in physics, Kant fails to give it a satisfactory answer. This failure is distilled in the equivocal status of the science of phoronomy. For it is here that the boundary between pure geometry and pure physics is crossed, without this being adequately acknowledged.

Let us briefly recall the crux of Kant's error. A central aim of the *Metaphysical Foundations* was to provide pure empirical constructions of the concepts of force and speed. In the case of force, a strict mathematical construction is particularly problematic since the category in question (that of cause and effect) is "dynamical." Unlike the notion of speed (*Geschwindigkeit*), our concept of force has no necessary connection to locomotion, or change in place (*Bewegung*). In a mathematised theory of nature, force must be explicitly characterised in terms of locomotion, in order that the requisite mathematical properties will result. But even at this point, there are still significant difficulties to be resolved. Speed may well be definitionally related to space (as change in place); however, speed is also an intensive magnitude, for speeds do not contain one another as parts. Although this concept can be "constructed" in a strict, mathematical sense, it still requires geometrical schemata. Once these have been provided, the additive characteristics of the concept of speed will be given, as Kant insists they must be, in terms of the congruence relations which form the subject of the properly apodictic science of space, that is to say they will be defined in terms of "geometric addition." And thus everything comes to depend on the status of the basic presuppositions of phoronomy itself: that the concept of an (unforced) motion is properly schematised as a straight line-segment whose length represents speed; that the laws of the composition of motions are given by geometrical constructions, where the line segments represent the speeds being combined. If this connection between basic kinematic and geometrical concepts is not explained and justified, then the reduction of the additive properties of the higher-level concepts in physical science will fail.

As I argued in Chapter 1, this reductive chain of definitions, when coupled to the requirement of positional determinacy, was used to prove the necessity of central forces, of the parallelogram law of forces, and indeed the redundancy of the law of inertia. The problem is that the very relations of congruence that Kant posited as being apodictically necessary cannot be apodictic in the sense required *once they are being applied in a phoronomic context.* For here we must, by definition, be dealing with empirically determinate magnitudes. And Kant cannot drop that claim at this stage in the argument without abandoning the principle of positional determinacy, which would mean invalidating the proofs of central forces, and so forth. For Newton, as we saw, this represented no special problem. He could, in a word, differentiate in an absolute space whose structure is described by Euclidean geometry. And he regarded his first law of motion as empirical. A realist attitude towards space, coupled with a generously empirical reading of his axioms renders his logic impeccable. However, the axioms of geometry apply to a part of the empirical world that is never experienced. Conversely, Kant's legitimate suspicion of the idea that geometry is a science of absolute space, combined with an excessive reliance on Leibnizian arguments (the repeated appeals to determinacy), results in a theory which is internally flawed, but nonetheless avoids introducing transcendent objects. Both Newton and Kant, in other words, are unable to explain how geometrical concepts actually determine physical appearances.

In Chapter 3, I argued that Helmholtz was saddled with the very same difficulties in Kant's theory. He too was aiming at an account of basic physical principles that made no appeals to absolute spatial relations. On his account as well, such appeals could be eliminated by assuming that all magnitudes relevant to physical theory were given by the physically determinate properties of an isolated system, in other words by the masses and relative positions of the particles composing it. These demands could be met by assuming force centrality, for in that case, no transcendent magnitudes had to be invoked in physics.[7] And this alone provided a transcendental justification for central-force theories: since the regulative demand imposed by the complete comprehensibility of nature would force us to reject non-central theories on these grounds, we are at least provisionally justified in constructing our physical principles on this model. Helmholtz was, however, frustrated in this project, because he was unable to demonstrate the equivalence of *vis viva* conservation and the central force hypothesis. In particular, he failed to show that if conservation held, then all forces were central. Had this implication held, it would have followed that all theories satisfying energy conservation could be formulated as central-force theories. Since the latter were epistemologically privileged, all theories involving non-central forces, above all electrodynamical theories which contained theoretical terms defined in terms of "something which could never be an object of a possible experience, namely … undifferentiated empty space" were to be rejected.[8]

The objections raised by Clausius, together with the colour-research described in Chapter 4, provided sufficient grounds to inquire more deeply into a concealed premise of these arguments. For the claim that relations to "undifferentiated empty space" are inadmissible, thus that all legitimate relations consist in the relations between real things, is hard to reconcile with the constitutive role of geometry in defining what Helmholtz had called general physical concepts in the Königsberger manuscript. The task he had set himself there was, from a philosophical point of view, virtually identical to that of Kant's *Metaphysical Foundations.* In this early manuscript, Helmholtz argues that certain principles are foundational in the natural sciences, and that these principles must be sought in the forms of all possible experience of the natural world. However, he has already broken with Kant when it comes to the relation between geometry and kinematics. Whereas Kant had argued that geometrical propositions were properties of all spatial intuitions, and that their proofs could be arrived at by means of constructions in pure intu-

7 *Cf.* (Helmholtz 1996), pp. 54–55. Even in these notes from 1882, Helmholtz insists on the importance of the principle that the forces within a system be known "once the position of the masses is completely given."

8 Again, *cf.* (Helmholtz 1996), pp. 54–55. The quotation is once again taken from the 1882 comments.

ition, Helmholtz introduced a refinement to this picture. Some properties of space are required in order for it to comprehend (*umfassen*) bodies in general; others are required for space to comprehend bodies in motion. Thus Helmholtz partitions Kant's theory of spatial intuition into motive and non-motive parts. In contrast to Kant's presentation, congruence relations belong to the motive part. This means that Helmholtz, well before he castigates Clausius for employing arbitrary coordinate systems in order to define non-central (but conservative) forces, already interprets statements concerning congruence relations as belonging to pure kinematics, and not to geometry proper.

In this early work, he is still unable to free himself of underlying definitions of length and distance, in terms of which he can define the concepts of a rigid body and thus of congruence relations. In contrast to the 1868 papers on geometry, the coordinate system used in providing the analytic definition is assumed from the outset to determine the length relations which go into the "motive" definition of congruence. From a modern point of view, in fusing classical geometry and kinematics by defining the concepts of both analytically, Helmholtz circumvents, but does not solve the real philosophical difficulty. Both sets of concepts are still derivative on the relations induced by the coordinate system, under the tacit assumption of a Euclidean metric. And thus Helmholtz is guilty in this early attempt of the same reliance on "purely mathematical" quantities as his later opponent Clausius was to be.

The first decisive break in Helmholtz's understanding of geometry is therefore the one that I described in Chapter 4. The importance of the research on colours lay in the fact that they satisfied the topological axioms of a continuous manifold, without constituting an *extensive* one. Because colours are intensive magnitudes, there is no natural or intuitive measuring procedure for quantifying them: they cannot be divided into parts, in order that units may be defined and compared with the wholes. Graßmann's analysis provided both Helmholtz and Maxwell with an alternative definition of what it means to be a magnitude: we can call something a magnitude when it satisfies certain axioms, among them axioms concerning transitivity and additivity, and when operational realisations of these measurement axioms are available. Helmholtz realised that by treating space itself as an *intensive* manifold, one could regard metrical properties as dependent on operations, and thus on relations defined exclusively among real things. This approach is characteristic of the subsequent papers on geometry, as well as the late work on measurement in *Zählen und Messen.* The guiding notion is that there should be what I might call *empirical closure* within natural science. The basic magnitudes involved in a theory should be specified by means of measurement axioms and metrological procedures. The notion of extensive spatial magnitudes, and thus of inherent geometrical properties, whether these be regarded as properties of spatial intuition, or of absolute space, is to be rejected. If the concept of an extensive magnitude is retained at all, then it can refer only to time.

Seen against this background, Helmholtz's strategy from 1870 onwards, namely from the point at which he realised the compossibility of Euclidean and non-Euclidean geometries, is a consistent development. For, as I suggested at the outset, the transcendental argumentation of the first two papers is not in fact anomalous. It represents a logical step in a series of efforts to render basic physical concepts empirically determinate by banishing references to absolute space. And these efforts were not merely philosophically motivated, for they also served the theoretical aims of the *Conservation* program. Given the strong Kantian bent of the Königsberger manuscript, we can characterise Helmholtz's strategy as follows: Kant, in the *Metaphysical Foundations*, had seen his task to consist in reductively defining the additive properties of intensive physical magnitudes in terms of the inherently additive properties of extensive magnitudes. The extensive structure of space and time—indeed the fact that they were the only such structures—provided a sufficient reason for claiming that all natural science had to be mechanistic. Only in this case would we be able to satisfy the demand that scientific concepts be rendered fully mathematical. The problem was that the extensive properties of physical *space* are not adequate to grounding the extensive relations required in empirical science. For what we need are not relations among regions of space, but among bodies in motion, that is to say among the measuring instruments used to establish distances along inertial paths.

Helmholtz, in treating spatial relations as if they were intensive magnitudes, effectively lopped off the last step in Kant's reduction. The fundamental magnitudes involved in physics were no longer geometrical properties in the classical sense. They were exactly those which Helmholtz had attempted to define in the Königsberger manuscript, without, however, succeeding in freeing himself from the notion of underlying and determinate metrical relations. For there, as we may recall, Helmholtz defined an inertial path as one whose direction did not change, and whose parts could be measured by treating any segment as a rigid body. Equally long parts of such an inertial path were therefore also congruent to one another. In this manner, the conceptual definition of an inertial path as a uniform and unchanging one is given a pure empirical construction. But this pure empirical construction is not in fact free of the metrical definition of length. As in Kant's phoronomy, the basic kinematic concepts, which for Helmholtz included congruence, were still defined in terms of geometric ones, albeit those of analytic geometry.

In the 1868 papers, this last step was effectively eliminated. By transcendentally proving that rigid bodies can only exist in a Euclidean space, Helmholtz offered a definition of the fundamental magnitudes that was physically realisable (so long as measurement was possible at all), and which did not depend on the assumption that there *were* unobservable metrical relations. As I suggested in Chapter 6, for Helmholtz a geometrical space is a manifold coupled to measurement axioms and a realisable measurement operation. If the operation can be realised, then there is no need to appeal to an underlying spatial metric. In claiming that physical concepts rest on empirically realisable measurement operations, Helmholtz had rendered physics empirically closed, in that he had eliminated any appeal to an inherent spatial metric. Secondly, he had shown that since it was at least logically possible that no such measurement procedure could be realised, one could regard the *possibility* of carrying out empirical measurements as contingent. Helmholtz would thereby have succeeded where Kant had failed, in that he would have rendered physical science capable—at least in principle—of providing a complete description of nature.

Even at the point where Helmholtz does come to insist on the inductive character of geometrical proposition, that is to say from 1870 onwards, this more basic claim persists, and indeed it is in some degree strengthened. Once Beltrami's work opened up the possibility of pseudo-spherical metrics, Helmholtz argued that the material basis of geometrical propositions could have led us to develop a different system of geometry. If the behaviours of everyday measuring instruments had exhibited a non-Euclidean behaviour, we would have been led to ascribe a non-Euclidean metric to space. In contrast to the opening of the 1868 paper, where Helmholtz deferred the question concerning the "origin of our knowledge," he now argues that this origin is to be found in our macroscopic, pre-scientific environment—what Husserl later came to call the life-world.[9] Nonetheless, as I argued in Chapter 6, these arguments in fa-

9 See for instance Husserl's "Die Frage nach dem Ursprung der Geometrie" (Husserl 1939), reprinted as an appendix to (Husserl 1962).

vour of an inductive empiricism do not detract from Helmholtz's continued emphasis on the relation between spatial magnitudes and the sciences. For geometry, considered as a formal science, does not merely describe the imperfect coincidence relations of everyday bodies. It remains a part of the natural sciences, in that it consists of axioms that "concern magnitudes."[10] And these axioms gain a factual content only when they are supplemented by basic natural laws, such as "the axiom of inertia or the single proposition that the mechanical and physical properties of bodies and their mutual reactions are ... independent of place."[11] This conception is given its final form in the 1878 reply to Land, where Helmholtz replaces congruence relations with the notion of "physical equivalence." The basic magnitudes of spatial science are now *explicitly* kinematic, or in Kant's terms phoronomic. Although there remains a certain conventional freedom open to us regarding which system of magnitudes is to be posited as basic, this freedom is tightly limited by the requirement that the resultant laws be position-independent. Indeed, the earlier definition resting on congruence is now seen only as a special case of the phoronomic definition, for the principle of positional independence is "of the utmost importance for the whole system of our mechanical and physical conceptions. That rigid solids, as we call them, which are really nothing else than elastic solids of great resistance, retain the same form in every part of space if no external force affects them, is a single case falling under the general principle."[12]

Now it is quite evident that Helmholtz has not gone the full distance. The definition of physically equivalent magnitudes which is foreshadowed in the above addition to the 1876 translation of the 1870 paper, still contains an appeal to notions such as "under the same circumstances" or "when no external force affects them." And thus, from the point of view of a strict conventionalist such as Poincaré, Helmholtz has ducked the question of what are to count as the same circumstances. As I pointed out when discussing Schlick's comments in Chapter 6, a proponent of a non-Euclidean metric would insist that what counted as physically equivalent situations in a Euclidean metric do not count as such in *his*. And thus the definition is equivocal. Helmholtz's response would be, on my reading, that whenever positionally dependent laws result from such a conventional choice, we have a systemic reason for rejecting them. It

10 *Cf.* (Helmholtz 1870), p. 29.
11 (Helmholtz 1876), p. 320.
12 (Helmholtz 1876), p. 320.

could have been the case that a non-Euclidean metric of constant curvature would yield positionally independent laws; however, given the empirical adequacy of the Newtonian-Euclidean system, we can show that this cannot be the case for the physical world in which we (or, better, Helmholtz) actually live. But both Poincaré and, later, Einstein,[13] concede the more fundamental point: whatever it is that geometric propositions describe, it is not merely the properties of space itself, but rather the properties of that system of bodies that we use to ascertain spatial, and indeed spatio-temporal equivalence-classes. If one wishes to insist on the separate existence of a pure *a priori* geometry, then the latter is certain only in so far as the logical connections among its propositions are valid. Once the latter is applied to physical reality, it becomes a subset of empirical science.

In conclusion, I shall say a few words concerning the significance of this conclusion from the point of view of the *Conservation* project, and thus its importance to Einstein's later notion of a "practical geometry." I have been arguing throughout this book that the interest in empiricising geometry was coupled from the outset to the idea that absolute space cannot be experienced, and thus that relations to absolute space are *a priori* inadmissible in physical theory. Helmholtz did not ever step down from this claim, indeed he reiterated it in the 1882 comments on the *Conservation* in a still stronger form. Although he now conceded that his attempts in the *Conservation* to reduce all action to central forces had failed, he still held to the basic philosophical principles that had driven that project. It might well be the case that the principle of decomposition that had been employed to derive centrality was indeed, as Newton had correctly surmised from the outset, an empirical and not an apodictic proposition. And yet, Helmholtz argues:

> Both this factual content of Newton's second axiom, as well as the proposition … that the forces holding among two bodies must necessarily be determined once the positions of the masses are known, have been abandoned in those electrodynamic theories which make the forces acting between two bodies depend on their speed and acceleration. Attempts made in this direction have always contradicted … the principle of action and reaction, and of the conservation of energy….

In other words, attempts to account for electromagnetic phenomena by means of non-central forces, although they cannot be ruled out on

13 In (Einstein 1921).

such *a priori* grounds as Helmholtz had initially offered, nevertheless violate basic physical principles. He continues:

> ... if a force is made dependent on absolute motion, that is to say on a changing relation between a mass and something that can never be an object of a possible experience, then this seems to me to be an assumption which abandons the hope of a complete solution of the task of natural science, which can only be allowed, in my opinion, once all other theoretical possibilities are exhausted.[14]

The touchstone of an epistemologically coherent system of physics continues to be that it provide a completely determinate system of nature. Furthermore, that system must be *empirically closed* in the sense outlined above: it cannot involve magnitudes that are positionally dependent, for instance determinations with respect to absolutes space. As Olivier Darrigol has emphasised, Helmholtz never truly abandons the principle of the complete determinacy of nature, nor is he prepared to relinquish the principle of decomposition, even if he now concedes that it is a contingent proposition.[15] Furthermore, as the second passage makes clear, those references to absolute spatial magnitudes which Helmholtz had sought to eliminate by borrowing Kant's arguments from what I called positional determinacy continue to be regulatively inadmissible.

These arguments are part and parcel of the same conception that underlies the project of rendering geometry phoronomic, indeed we see Helmholtz reiterating them here in 1882, only a few years after the geometric research programme is given its final form in the 1878 reply to Land. When we combine both lines of argument, we see that what Helmholtz is requiring is a theory of electrodynamics that does not involve motion with respect to the ether, or indeed any basic magnitudes that are not representable as point-coincidences. Whatever forces and motions are admitted into physics, it must be at least possible to imagine them as determined by observable properties of point-systems. Of course, "observable" is being used here in a rather rarefied sense: what is meant is more accurately captured by Kant's notion of pure empirical construction. We must be able to imagine the world as consisting of point-masses whose relations among one another are sufficient to determine the application of general laws to them. Given the empirical critique of geometry offered by Helmholtz, we must also regard the measurement devices employed as part of the systems being described. Furthermore, the application of mathematics

14 (Helmholtz 1996), pp. 54–55.
15 (Darrigol 1994), p. 222.

made in such descriptions cannot be justified, as it was in Kant, by assuming that space is, in itself, a magnitude. On the contrary, the notion of a magnitude must itself be explained in terms of measurement operations based on coincidence relations. If it is true, as Gregor Schiemann has argued,[16] that Helmholtz's commitments are those of a mechanist, then they are so by reason of his epistemological convictions.

This strongly phenomenological viewpoint may seem almost atavistic when advanced in 1882; however, it is the same model adopted by Weyl, Einstein, Schlick and many others in the following years, when the problems raised by Maxwell's theory—a theory Helmholtz never quite accepted—become truly inescapable.[17] Why did all these authors feel compelled to follow Helmholtz in his investigations? According to the conventional reading of Helmholtz against which I have been arguing, the work on geometry provided a useful tool for addressing the problems raised by the concept of absolute space once the Michelson-Morley results were in. But it is far more accurate, on my view, to interpret Helmholtz as engaging in these investigations because he was already concerned with this *physical* problem. Furthermore, the direction taken by his analysis must be seen as resulting from a commitment to essential aspects of Kant's philosophy of science, and not solely as a reaction against the latter's epistemology of mathematics. German scientists such as Helmholtz and Mach were far more sceptical than their British counterparts of the notion of absolute space, and of its role in Newtonian mechanics, and this scepticism is in large measure due to two of Kant's doctrines that we have discussed

16 (Schiemann 1997). Schiemann is opposed to reading the *Conservation* from a Kantian point of view, p. 189. But while I agree with his criticisms of (Heimann 1974), I obviously do not agree with his suggestion that no significant relation exists between the two authors. Indeed, I see no alternative to viewing Helmholtz's commitment to mechanism as primarily an epistemological one.

17 Einstein observes (Einstein 1921) that the concept of an empirical geometry based on rigid-body motion was essential to his development of *general* relativity, and thus it is not the problem of motion relative to the ether that makes this concept useful to him. Nevertheless, it is the solution that Einstein gives to the problem of absolute motion which first engenders the conceptual difficulties that induced him to reanimate Helmholtz's idea. Once special relativity has disposed of absolute motion and the problems generated by electrodynamics by means of a kinematic interpretation of the Lorentz contractions, non-Euclidean behaviours of rigid bodies arise. Since these bodies are, as in Helmholtz, logically coupled to the idea of an inertial frame, Einstein can use this insight to explore the conditions of gravitational and inertial equivalence. *Cf.* (Friedman 2002), pp. 211–212.

at length in this book: the doctrine of space and time as forms of possible experience; and the doctrine that whatever is physically meaningful must refer to a possible appearance. These doctrines result in an epistemological commitment to a strong form of spatial relativity which barred admission of any absolute frame of reference—a point of view clearly expressed in the above quotation.

This commitment to relativism contains the seeds of the problem of space, and of the solution that Helmholtz proposed to it. Mach, writing at the turn of the century, expressed the situation as follows:

> The view that "absolute motion" is a conception which is devoid of content and cannot be used in science struck almost everybody as strange thirty years ago, but at the present time it is supported by many worthy investigators. … Probably there will soon be no important supporters of the opposite view. But, if the inconceivable hypotheses of absolute space and absolute time cannot be accepted, the question arises: In what way can we give a comprehensible meaning to the law of inertia?[18]

My suggestion has been that Helmholtz was one of the first people to pose himself Mach's question. Since the law of inertia makes explicit assertions about the behaviours of bodies, and since these behaviours are singled out by their satisfying certain geometrical definitions, it is here that geometric propositions first gain an empirical meaning (at least in Newtonian mechanics). But this means that for a relativist, there is no way of giving the law a meaning—it does not express a proposition that could be true or false. Kant induces this conflict between the requirement of empirical determinacy and the doctrine of spatial relativity; however, he overlooks its significance. He thinks that the law of inertia is superfluous, and at that point in his system where geometry becomes an empirical science, namely in phoronomy, he demands that kinematic propositions be rendered geometric, but he does not adequately explain how this can be done.

Thus it is Helmholtz who, in borrowing these arguments, and in attempting to state his law of energy conservation without any reference to relations that are not determined by the bodies in a system, first truly confronts this difficulty. His hope, as is reflected in both the Königsberger manuscript and the methodology of the 1868 papers is to give an *a priori* foundation to the notion of what he later called physically equivalent magnitudes. That is to say, he wishes to leapfrog pure geometry and get straight to the foundational magnitudes of *physical* space. For

18 (Mach 1960), p. 293

in such a case, principles such as that of the conservation of energy or the law of inertia would no longer contain an empirically indeterminate, geometric part. Helmholtz's remarks in the reply to Clausius concerning the necessity of using only empirically determinate coordinate systems are still only targeted at the notion of an absolute direction. However, the attempt to establish a kinematic geometry in the Königsberger manuscript antedates these remarks, just as the work on the metric of the colour-space is simultaneous to them. Thus we can fairly conclude that the first geometry papers were intended to complete this project of empiricising the science of space by fusing phoronomy with geometry. From the point of view of the trend that Mach describes in the above quotation, Helmholtz's work no longer appears as an investigation of problems in the epistemology of mathematics which then proved, by reason of fortuitous timing, to be relevant to physical problems. Nor is his subsequent introduction of physically equivalent magnitudes a mere corrective to his earlier approach. We should instead view Helmholtz's research programme as the result of a sustained exploration of the hidden difficulties in Kant's theories of space and motion. The relevance of these investigations to the work of Poincaré and Einstein is then not an accidental one. It is rather the natural development of a process set in motion, as Mach observed, thirty years before.[19]

19 (Mach 1960), p. 293.

Bibliography

Arthur, Richard. 2009 forthcoming. On Newton's Fluxional Proof of the Vector Addition of Motive Forces. In *Infinitesimals*, edited by W. Harper and C. Fraser.

Beutelspacher, Albrecht. 1996. A Survey of Graßmann's *lineale Ausdehnungslehre.* In *Hermann Günther Graßmann (1809–1977): Visionary Mathematician, Scientist, and Neohumanist Scholar*, edited by D. Schubring. Dordrecht: Kluwer.

Bevilacqua, Fabio. 1993. Helmholtz's Über die Erhaltung der Kraft: The Emergence of a Theoretical Physicist. In *Hermann von Helmholtz and the Foundations of Nineteenth-century Science*, edited by D. Cahan. Berkeley: Univ. of California Press.

———. 1994. Theoretical and Mathematical Interpretations of Energy Conservation: The Helmholtz-Clausius Debate on Central Forces 1852–54. In *Universalgenie Helmholtz*, edited by L. Krüger. Berlin: Akademie Verlag.

Carnap, Rudolf. 1922. *Der Raum: ein Beitrag zur Wissenschaftslehre.* Berlin: Reuther & Reichard.

Carrier, Martin. 1994. Geometric Facts and Geometric Theory: Helmholtz and the 20th Century Philosophy of Physical Geometry. In *Universalgenie Helmholtz*, edited by L. Krüger. Berlin: Akademie Verlag.

Clausius, Rudolph. 1853. Über einige Stellen der Schrift von Helmholtz Über die Erhaltung der Kraft. *Annalen der Physik* 89:568–579.

———. 1854. Über einige Stellen der Schrift von Helmholtz Über die Erhaltung der Kraft; zweite Notiz. *Annalen der Physik* 91:601–604.

Cohen, B., and K. Westfall, eds. 1995. *Newton: Texts, Backgrounds, Commentaries.* New York: Norton.

Darrigol, Olivier. 1994. Helmholtz's Electrodynamics and the Comprehensibility of Nature. In *Universalgenie Helmholtz*, edited by L. Krüger. Berlin: Akademie Verlag.

———. 2003. Number and measure: Hermann von Helmholtz at the crossroads of mathematics, physics, and psychology. *Studies in the History and Philosophy of Science* (34):515–573.

DiSalle, Robert. 1993. Helmholtz's Empiricist Conception of Mathematics: Between Laws of Perception and Laws of Nature. In *Hermann von Helmholtz and the Foundations of Nineteenth-century Science*, edited by D. Cahan. Berkeley: Univ. of California Press.

———. 2008. "Space and Time: Inertial Frames", The Stanford Encyclopedia of Philosophy (Fall 2008 Edition), Edward N. Zalta (ed.), URL = <http://plato.stanford.edu/archives/fall2008/entries/spacetime-iframes/>.

Dubucs, Jacques. 1998. Beth, Kant et l'intuition mathématique. *Philosophia Scientiae* 3 (4).

Einstein, Albert. 1921. *Geometrie und Erfahrung. Erweiterte Fassung.* Berlin: Springer.

Foucault, Léon. 1853a. La recomposition des couleurs du spectre en teintes plates. *Cosmos* (II):232.

———. 1853b. Über die Wiedervereinigung der Strahlen des Spectrums zu gleichförmigen Farben. *Pöggelers Annalen* 88:385–387.

———. 1878. *Recueil des travaux scientifiques de Léon Foucault.* Paris: Gauthier-Villars.

Friedman, Michael. 1986. The Metaphysical Foundations of Newtonian Science. In *Kant's Philosophy of Physical Science*, edited by R. Butts. Dordrecht: Reidel.

———. 2002. Geometry as a Branch of Physics: Background and Context for Einstein's "Geometry and Experience". In *Reading Natural Philosophy*, edited by D. B. Malament. Chicago: Open Court.

Graßmann, Hermann. 1853. Zur Theorie der Farbenmischung. *Pöggelers Annalen* 89:69–84.

———. 1877. Bemerkungen zur Theorie der Farbenempfindungen. In *Gesammelte Werke*, edited by F. Engel. Leipzig: Teubner.

———. 1878. *Die lineale Ausdehnungslehre.* Leipzig: Otto Wigand.

Hatfield, Gary. 1990. *The Natural and the Normative. Theories of Spatial Perception from Kant to Helmholtz.* Cambridge, Mass.: MIT Press.

Heidelberger, Michael. 1993. Force, Law and Experiment: The Evolution of Helmholtz's Philosophy of Science. In *Hermann von Helmholtz and the Foundations of Nineteenth-century Science*, edited by D. Cahan. Berkeley: Univ. of California Press.

Heimann, P.M. 1974. Helmholtz and Kant: The Metaphysical Foundations of *Über die Erhaltung der Kraft. Studies in the History and Philosophy of Science* 5:205 ff.

Helmholtz, Hermann von. 1852a. Über die Theorie der zusammengesetzten Farben. In *Wissenschaftliche Abhandlungen, 2. Bd.* Leipzig: Johann Ambrosius Barth.

———. 1852b. Über Herrn D. Brewster's neue Analyse des Sonnenlichtes. In *Wissenschaftliche Abhandlungen. 2. Bd.* Leipzig: Johann Ambrosius Barth.

———. 1852c. Über die Natur der menschlichen Sinnesempfindungen. In *Wissenschaftliche Abhandlungen. 2. Bd.* Leipzig: Johann Ambrosius Barth.

———. 1854. Erwiderung auf die Bemerkungen von Hrn. Clausius. In *Wissenschaftliche Abhandlungen. 1. Bd.* Leipzig: Johann Ambrosius Barth.

———. 1855. Über die Zusammensetztung von Spectralfarben. In *Wissenschaftliche Abhandlungen, 2. Bd.* Leipzig: Johann Ambrosius Barth.

———. 1863a. Über die Bewegungen des menschlichen Auges. In *Wissenschaftliche Abhandlungen. 2. Bd.* Leipzig: Johann Ambrosius Barth.

———. 1863b. Über die normalen Bewegungen des menschlichen Auges. In *Wissenschaftliche Abhandlungen. 2. Bd.* Leipzig: Johann Ambrosius Barth.

———. 1867. *Handbuch der Physiologischen Optik.* 1 ed. Leipzig: Leopold Voss.

———. 1868a. Über die tatsächlichen Grundlagen der Geometrie. In *Wissenschaftliche Abhandlungen. 2. Bd.* Leipzig: Johann Ambrosius Barth.

———. 1868b. Über die Tatsachen, die der Geometrie zum Grunde Liegen. In *Wissenschaftliche Abhandlungen. 2. Bd.* Leipzig: Johann Ambrosius Barth.
———. 1870. Über den Ursprung und Bedeutung der geometrischen Axiome. In *Vorträge und Reden. 2. Bd.* Leipzig: Johann Ambrosius Barth.
———. 1872. Über die Theorie der Elektrodynamik. Vorläufiger Bericht. In *Wissenschaftliche Abhandlungen. 1. Bd.* Leipzig: Johann Ambrosius Barth.
———. 1876. The Origin and Meaning of Geometrical Axioms (I). *Mind* 2:301–321.
———. 1878a. Die Tatsachen in der Wahrnehmung. In *Vorträge und Reden. 2. Bd.* Braunschweig: Vieweg.
———. 1878b. The Origin and Meaning of Geometrical Axioms (II). *Mind* 10:212–225.
———. 1878c. Über den Ursprung und Sinn der geometrischen Sätze; Antwort gegen Herrn Professor Land. In *Wissenschaftliche Abhandlungen. 2. Bd.* Leipzig: Johann Ambrosius Barth.
———. 1882. *Wissenschaftliche Abhandlungen.* 3 vols. Vol. 1. Leipzig: Johann Ambrosius Barth.
———. 1887. Zählen und Messen, erkenntnistheoretisch betrachtet. In *Schriften zur Erkenntnistheorie*, edited by P. Hertz a. M. Schlick. Berlin: Springer.
———. 1891a. Kürzeste Linien im Farbensystem. In *Wissenschaftliche Abhandlungen. 3. Bd.* Leipzig: Johann Ambrosius Barth.
———. 1891b. Versuch, das psychophysische Gesetz auf die Farbenunterschiede trichromatischer Augen anzuwenden. In *Wissenschaftliche Abhandlungen. 3. Bd.* Leipzig: Johann Ambrosius Barth.
———. 1891c. Versuch einer erweiterten Anwendung des Fechner'schen Gesetzes im Farbensystem. In *Wissenschaftliche Abhandlungen. 3. Bd.* Leipzig: Johann Ambrosius Barth.
———. 1896. *Handbuch der Physiologischen Optik.* 2 ed. Hamburg and Leipzig: Leopold Voss.
———. 1911. *Handbuch der Physiologischen Optik.* 3 ed. 3 vols. Hamburg and Leipzig: Leopold Voss.
———. 1921. *Schriften zur Erkenntnistheorie.* Edited by P. Hertz a. M. Schlick. Berlin: Springer.
———. 1977. *Epistemological Writings: the Paul Hertz/Moritz Schlick Centenary Edition of 1921.* Translated by M. F. Lowe. Edited by R. S. Cohen and Y. Elkana, *Boston Studies in the Philosophy of Science.* Dordrecht: Reidel.
———. 1996. *Über die Erhaltung der Kraft [u.a.].* Frankfurt a.m.: Harri Deutsch.
———. 1998. *Schriften zur Erkenntnistheorie.* Edited by P. Hertz a. M. Schlick. Vol. 2, *Kleine Bibliothek für das 21. Jahrhundert.* Wien: Springer.
Husserl, Edmund. 1939. Die Frage nach dem Ursprung der Geometrie als intentional-historisches Problem. *Revue internationale de philosophie* (2):203–225.
———. 1962. *Die Krisis der europäischen Wissenschaften und die transzendentale Phänomenologie.* Vol. 6, *Husserliana.* The Hague: Martinus Nijhoff.
Hyder, David. 1999. Helmholtz's Naturalized Conception of Geometry and his Spatial Theory of Signs. *Philosophy of Science* (66 (Proceedings)):S273-S286.

———. 2003. Kantian Metaphysics and Hertzian Mechanics. In *The Vienna Circle and Logical Empiricism*, edited by F. Stadler. Dordrecht: Kluwer.

Kant, Immanuel. 1910. *Gedanken von der wahren Schätzung der lebendigen Kräfte*. In *Kants Werke. Bd 1*. Berlin: Preußiche Akademie der Wissenschaften.

———. 1911a. *Metaphysische Anfangsgründe der Naturwissenschaft*. In *Kants Werke Bd. 4*. Berlin: Preußiche Akademie der Wissenschaften.

———. 1911b. *Prolegomena zu einer jeden künftigen Metaphysik*. In *Kants Werke Bd. 4*. Berlin: Preußiche Akademie der Wissenschaften.

———. 1911c. *Jäschke Logik*. In *Kants Werke Bd. 9*. Berlin: Preußiche Akademie der Wissenschaften.

———. 1998. *Kritik der reinen Vernunft*. Vol. 505, *Philosophische Bibliothek*. Hamburg: Meiner.

Kremer, Richard L. 1993. Innovation through Synthesis. Helmholtz and Color Research. In *Hermann von Helmholtz and the Foundations of Nineteenth-century Science*, edited by D. Cahan. Berkeley: Univ. of California Press.

Kries, Johannes von. 1882. Über die Messung intensiver Größen und über das sogenannte psychophysiche Gesetz. *Vierteljahrschrift für wissenschaftliche Philosophie* (6):257–294.

Königsberger, Leo. 1903. *Hermann von Helmholtz*. 3 vols. Vol. 2. Braunschweig: Vieweg.

Land, J.P.N. 1877. Kant's Space and Modern Mathematics. *Mind* 2 (5):38–46.

Lenoir, Timothy. 1993. The Eye as Mathematician: Clinical Practice, Instrumentation, and Helmholtz's Construction of an Empiricist Theory of Vision. In *Hermann von Helmholtz and the Foundations of Nineteenth-century Science*, edited by D. Cahan. Berkeley: Univ. of California Press.

Mach, Ernst. 1960. *The Science of Mechanics*. Translated by J. McCormack. 6 ed. LaSalle: Open Court.

Maxwell, James Clerk. 1855. Experiments on colour, as perceived by the eye, with remarks on colour-blindness. *Transactions of the Royal Society of Edinburgh* 21 (2):275–298.

Newton. 1931. *Opticks*. London: G. Bell and Sons.

———. 1968. *The Mathematical Principles of Natural Philosophy*. Translated by A. Motte. Vol. 1. London: Dawsons.

Poincaré, Henri. 1909. *La Science et l'hypothèse*, *Bibliothèque de philosophie scientifique*. Paris: Flammarion.

Pollok, Konstantin. 2001. *Kants Metaphysische Anfangsgründe der Naturwissenschaft. Ein Kritischer Kommentar*, *Kant-Forschungen 13*. Hamburg: Meiner.

Riemann, Bernhard. 1854. Über die Hypothesen, welche der Geometrie zu Grunde liegen. In *Berhard Riemanns gesammelte mathematische Werke*, edited by H. Weber. Leipzig: Teubner.

Russell, Bertrand. 1996. *An Essay on the Foundations of Geometry*. London: Routledge.

Schiemann, Gregor. 1997. *Wahrheitsgewissheitsverlust. Hermann von Helmholtz' Mechanismus im Anbruch der Moderne*. Darmstadt: Wissenschaftliche Buchgesellschaft.

Sherman, Paul D. 1981. *Colour Vision in the Nineteenth Century.* Bristol: Adam Hilger.

Sutherland, Daniel. 2004. The Role of Magnitude in Kant's Critical Philosophy. *Canadian Journal of Philosophy* 34 (3):411–442.

———. 2005. Kant on Fundamental Geometrical Relations. *Archiv für Geschichte der Philosophie* 87 (2):117–158.

Torretti, Roberto. 1978. *Philosophy of Geometry from Riemann to Poincaré.* Dordrecht: Reidel.

Turner, R. Steven. 1994. *In the Eye's Mind: Vision and the Helmholtz-Hering Controversy.* Princeton, NJ: Princeton Univ. Press.

Wahsner, Renata. 1994. Funktion und aposteriorische Herkunft: Hermann von Helmholtz' Untersuchungen zum Erfahrungsstatus der Geometrie. In *Universalgenie Helmholtz*, edited by L. Krüger. Berlin: Akademie Verlag.

Weyl, Hermann. 1927. *Philosophie der Mathematik und Naturwissenschaft.* Munich: Oldenbourg.

Index

www.ingramcontent.com/pod-product-compliance
Lightning Source LLC
LaVergne TN
LVHW010859110826
845149LV00005B/1424
9783110481570